HISTORY OF PLACER AND QUARTZ GOLD MINING IN THE COEUR D'ALENE DISTRICT

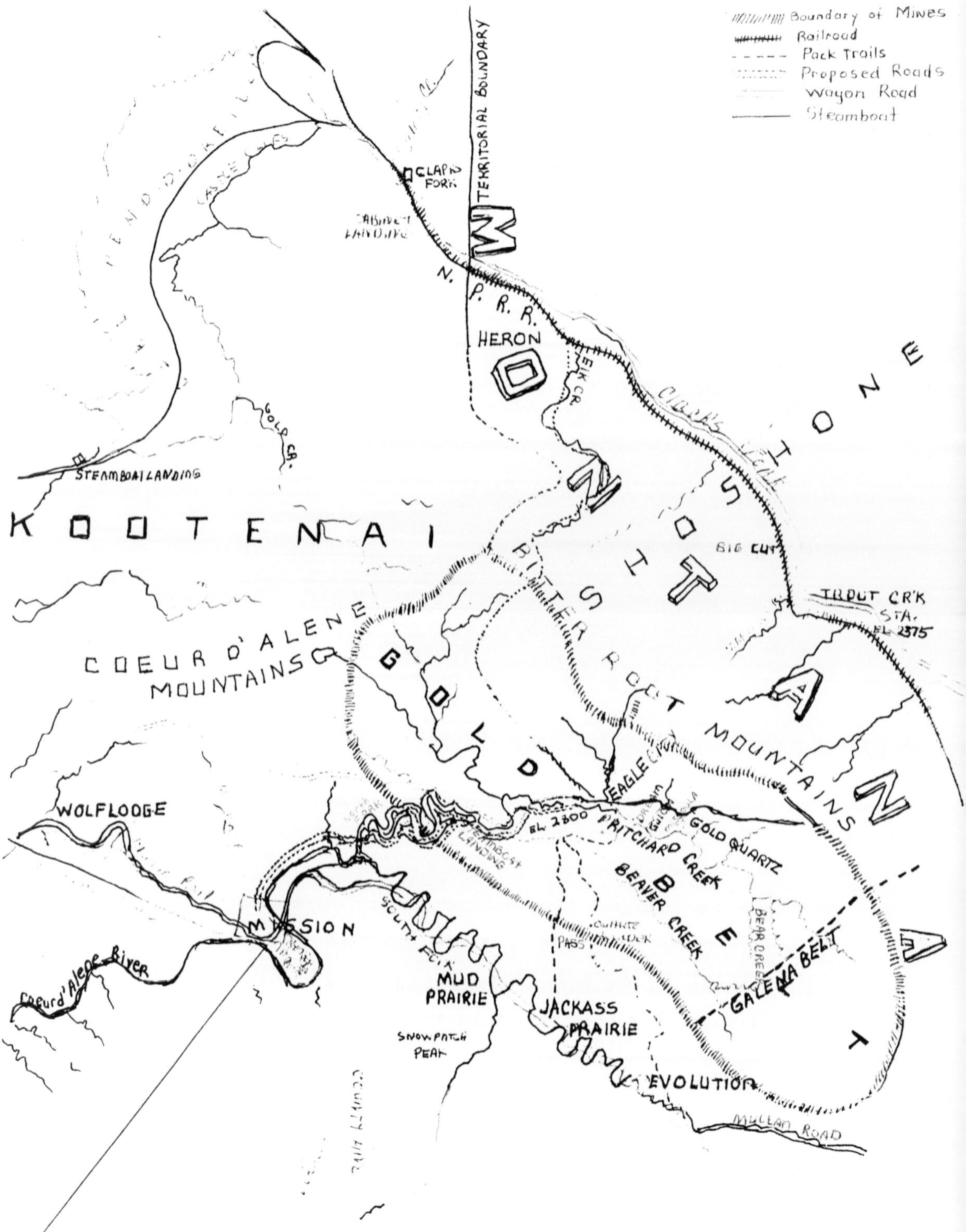

Boundary of Mines
Railroad
Pack Trails
Proposed Roads
Wagon Road
Steamboat
TERRITORIAL BOUNDARY
CLARK FORK
N. P. R. R.
HERON
MONTANA
YELLOWSTONE
BIG CUT
TROUT CR'K STA. EL 2375
STEAMBOAT LANDING
GOLD CR.
KOOTENAI
COEUR D'ALENE MOUNTAINS
BITTER ROOT MOUNTAINS
GOLD BELT
EAGLE CR.
EL 2800
PRITCHARD CREEK
GOLD QUARTZ
BEAVER CREEK
BEAR CREEK
GALENA BELT
WOLFLODGE
MISSION
SOUTH FORK
Coeurd'Alene River
MUD PRAIRIE
JACKASS PRAIRIE
SNOWPATCH PEAK
PASS
EVOLUTION
MULLAN ROAD

HISTORY OF PLACER AND QUARTZ GOLD MINING IN THE COEUR D'ALENE DISTRICT

A Thesis

Presented in partial fulfilment of the requirements for the
Degree of Master of Arts in the
Department of American History
of the
University of Idaho

By

Robert Wayne Smith
1932

YE GALLEON PRESS
FAIRFIELD, WASHINGTON

Library of Congress Cataloging-in-Publication Data

Smith, Robert Wayne, 1903-

History of placer and quartz gold mining in the Coeur d'Alene District / by Robert Wayne Smith.

p. cm.

Originally present as the author's thesis (Masters) — University of Idaho. 1932.

Includes bibliographical references (p.) and index.

ISBN 0-87770-519-4

1. Gold mines and mining — Idaho — Coeur d'Alene — History. 2. Gold mines and mining — Idaho — Murray — History. 3. Hydraulic mining — Idaho — Coeur d'Alene — History. 4. Hydraulic mining — Idaho — Murray — History. I. Title.

TN423.I2S63 1993 622'.3422'0979691 — dc20 93-7321

Prefatory Note

This historical study is the fruit of two years of research to satisfy a persistent curiosity. When I first came to Wallace, Idaho, in the fall of 1929, I began to inquire of the citizens the early history of the community. Everyone spoke very confidently of a gold mining boom 'way back in the 1880's, but no one had anything beyond hazy generalizations regarding that episode.

I became extremely curious about the gold country. I began to read historical references and also to search out "old-timers" who knew the story at first hand. These activities involved me in two summers of researching in the library of the University of Idaho at Moscow, Idaho, and a long camping and hiking trip to Murray, Idaho, and its vicinity. The latter project was followed by many short trips to the scenes of the early mining enterprise.

In the course of these activities, I established many hitherto unpublished facts, and incidentally I satisfied my curiosity. I also came in contact with a host of charming and patient people who gave me assistance.

In consideration of the facts determined, I present this thesis to the University of Idaho in the hope of adding a worthy chapter to the recorded history of this state.

With the further desire of sharing the personal satisfaction derived from this research, I also present this thesis to the reader, hoping that he, too, may find interest and pleasure in a narrative of human experience from a by-gone day.

Robert Wayne Smith.

Wallace, Idaho. November 20, 1932.

Acknowledgments

I take this opportunity to acknowledge publicly and with gratitude some services which were rendered to me in the course of this study.

First of all, I am indebted to Dr. Cornelius J. Brosnan, of the American History department at the University of Idaho, who directed my research and enabled me to avoid the pitfalls of inexperience. Also to the library staff of the University of Idaho who gave me special attention and special privileges in my reference work.

I wish to thank the Idaho State Historical Society at Boise, Idaho, for their generosity in allowing me to borrow valuable old newspapers from their files.

To Messrs. Jerome J. Day, Harry L. Day, and Harry Marsh, of Wallace, Idaho, I am also indebted. These gentlemen extended to me the use of their newspaper files and also made photostatic copies of old newspapers for me.

I owe a great deal to the kindness and patience of Mr. Adam Aulbach, of Murray, Idaho, who extended to me the use of his personal library and his private scrap-book of newspaper clippings. Mr. Aulbach spent many hours explaining in great detail the various incidents of which he was a direct observer.

To Mr. Howard Staples, of Moscow, Idaho, I am indebted for the majority of my information regarding the dredging work of the Yukon Gold Company.

The large photographs which illustrate this thesis came from the old plates at the Barnard's Studio, Wallace, Idaho, and they were located and printed for me by Miss Nellie Stockbridge, proprietor of the studio.

LIST OF ILLUSTRATIONS AND MAPS

TABLE OF CONTENTS

CHAPTER I

Into The Coeur D'Alenes, 1932[1]

At the very beginning of this historical sketch I should like to take the reader upon an imaginary tour into one of the most interesting regions of the Pacific Northwest.

Let us climb aboard a swift and powerful automobile at Coeur d'Alene, Idaho. We glide smoothly eastward on Sherman Avenue, the main thoroughfare of this attractive little city. At the eastern edge of town a shield-shaped road sign announces United States Highway Number 10, and beckons us to turn to the right. For two miles now we travel a smooth concrete highway paralleling the shores of Coeur d'Alene Lake. Then the road commences to climb, and in the next twelve miles we ascend two summits which afford us magnificent views fo the blue-green waters of the lake and the forested hillsides hemming it in.

At Wolf Lodge Creek the climb begins in earnest. The grade increases. Our car ascends in sweeping curves the Fourth of July Canyon, well known to motorists throughout the Inland Empire. Toward the top of the canyon the grade steepens noticeably; the curves become sharper. We pause for relaxation and perhaps for refreshment as we dismount at the heavy log tavern at the summit. As we look about we cannot fail to see a concrete monument ar the roadside bearing the figure of Captain John Mullan with the dates 1853–1855. A green lettered sign also announces that we are in the Coeur d'Alene National Forest, and that the altitude is 3160 feet above sea level.[2]

Resuming our journey we begin to descend the eastern grade into the valley of the Coeur d'Alene River, into the historic Coeur d'Alene district known throughout the mining world for its enormous mineral wealth. We have not gone far from the summit when our attention is attracted by a modest little sign which reads "Mullan Tree" and points with an arrow down a narrow forest path. Let us dismount again and follow the winding path. There in the forest stands a tall white pine tree surrounded by an iron railing. About shoulder high on the trunk is seen the broad axe-blaze cut by Captain John Mullan on a summer day in 1862.

Ten or twelve miles beyond the Mullan Tree, when our motor car has returned to the valley level, we sight a cluster of white buildings on the top of a grassy knoll a mile to the right of the highway. Let us turn off here and follow a narrow and perhaps muddy road across the meadows to this group of ancient buildings. The central structure is the Mission of the Sacred Heart, known locally as the Cataldo Mission, whose history goes back to 1847, although the mission building was not completed until 1868. The old mission has been well preserved, and its high whitewashed facade stands as the most venerable and most revered of all the historic monuments in the Coeur d'Alene district.

Back on the highway again we meet and cross the Coeur d'Alene River at Cataldo. The river, about 60 feet wide, flows a turgid current of greenish-gray color, unlike any other mountain stream. "What a dirty river!" you exclaim in disappointment and disgust. Yet this murky-looking water is but the first indication of the great mining activity of the district, for it is filled and colored by finely pulverized rock particles from the concentrating mills which prepare the lead and silver and zinc ores for smelting.

Signs of mining activity multiply as we proceed eastward. Nine miles beyond Cataldo we pass the mouth of Government Gulch. If we are not driving too fast on the excellent pavement we may catch a glimpse up the gulch of the tall smoke stack of the Bunker Hill and Sullivan Mining and Concentrating Company's new electrolytic zinc refining plant. The zinc plant, one of the most modern of its kind in the world, is well worth a visit, but we hurry on. Within another mile we have passed the Bunker Hill and Sullivan smelter and the same company's concentrating mill, the latter at the western city limits of Kellogg. These vast enterprises present a bewildering maze of buildings large and small, water flumes, smoke stacks, railroad tracks, electric wires, piles of timbers and other mysterious looking equipment on both sides of the highway and overhead.

We drive cautiously through Kellogg, a bustling city whose economic life centers upon the activities fo the Bunker Hill and Sullivan Company, one of the world's greatest producers of lead and silver. We are now in the heart of the Coeur d'Alene district. Every hill which rises above the river valley shows a sprinkling of mining "prospects;" i.e., a tunnel into the hill and a dump of loose rock below the tunnel on the hillside. These hills now rise steeply above the river, leaving in most places only a narrow valley or canyon through which run highway, railroad, and river—the South Fork of the Coeur d'Alene River, now a turbulent mountain creek of the same muddy gray color.

Twelve miles east of Kellogg we enter Wallace, the county seat of Shoshone County, Idaho, a beautiful residential city with clean, wide, paved streets. The city is situated on a level mountain meadow, surrounded on all sides by high hills. Although we see no great mines at Wallace, we sense at once in every conversation with its citizens an absorbing interest in mining enterprises. And indeed, an accurate gauge of the prosperity of this community can be quickly obtained by quoting the current prices of lead, silver, and zinc.

At Wallace our touring party is confronted with a choice of possible routes by which to continue the visit of the district. We might proceed on up the main valley, still on United States Highway Number 10, to Mullan and on across the Bitter Root divide into Montana, in which case we should see the great development of the Morning mine at Mullan, a property of the Federal Mining and Smelting Company. Or we might turn to the left and ascend Canyon Creek to Burke, a drive of only seven miles in which we should climb a thousand feet in altitude through a canyon lined with mines and mills both old and new, and crowned with the impressive works of the Hecla Mining Company at Burke.

But, if I am still the conductor or guide of the party, I should order at Wallace a still sharper turn to the left, into Nine-mile canyon, and our car would begin a beautiful twenty-mile drive over two mountain ridges into a corner of the Coeur d'Alene district almost forgotten by the outside world, which holds a story even more romantic and fascinating than the successful and prosperous mining communities through which we have already passed. Over a winding road, much narrower than the highway, yet safe withal, we climb up Nine-mile Canyon to Dobson Pass. Down the north slope of the ridge through a fine forest and we are on Beaver Creek down whose ever-widening valley we proceed. About 14 miles from Wallace we come upon a cross-roads sign marked Delta, and here we turn to the right up the valley of Trail Creek. We cross the second summit at King Pass where we enter the valley of Prichard Creek. Far down below us covering all the bottom of the valley lies a great pile of boulders, a feature which distinguishes this valley from others. As we descend the last stage of the road, the boulders come to be the most prominent feature of the landscape. And finally, after rounding a wide curve, the road cuts across the pile of boulders, crosses Prichard Creek, and carries us into Murray, Idaho. Rows of old frame buildings, many of them empty and dilapidated, wooden sidewalks in bad state of repair, a horse or a cow meandering leisurely down the street, a fat and contented cat sunning herself on a doorstep—these are the characteristics of Murray which first strike the eye of the visitor.

We have entered a strange and different world. We find ourselves transported back to the 1890's, or even the 1880's. Our modern automobile seems out of place here. Let us abandon it, and inspect this old town in pedestrian fashion. Here is a cool and shady corner beside an old hotel building. Sit down and let me reconstruct for you the story of this and other similar communities of the North Side of the Coeur d'Alene district.[3]

1. The term Coeur d'Alenes, in general use throughout the Pacific Northwest, refers to the mountainous area of Shoshone County, Idaho, drained by the Coeur d'Alene River.

2. A tunnel has just been opened, July, 1932, which carries the highway on a lower level over the ridge—or rather, under it. At present writing, neither the Mullan monument nor the forestry signs have been located on the new road.

3. The term North Side is a local phrase which designates the region drained by the North Fork of the Coeur d'Alene River and its tributaries. The parallel term South Side refers to the drainage area of the South Fork of the Coeur d'Alene River. A brief study of the map of the district will clarify the two Sides.

CHAPTER II

Fifty Years Ago

In 1880 the Coeur d'Alenes presented a far different appearance from that described in the previous chapter. There was no concrete highway to carry a speeding automobile. Instead, a narrow wagon road threaded a tortuous passage through a densely forested valley.

Captain John Mullan, (then Lieutenant) of the United States Army laid out the route and supervised the building of this road in the years 1859 to 1862. It was a military project whose terminals were Fort Walla Walla on the Columbia and Fort Benton on the Missouri. The Mullan Road followed the same general course as the present United States Highway Number 10 from Fort Coeur d'Alene to the Old Mission at Cataldo over the Fourth of July summit; and thence up the South Fork of the Coeur d'Alene River, crossing the Bitter Root divide through the Sohon or St. Regis Pass. Captain Mullan's road through the Coeur d'Alenes was narrow and steep, so narrow that wagons could scarcely pass between the trees on either side, and so steep that the worthy captain himself suggests that double teams of horses would probably be required to pass the summit (Sohon Pass). There were few habitations along the road, and most of these few were primitive cabins.

The most pretentious establishment in the Coeur d'Alene valley was the Old Mission of the Sacred Heart at Cataldo. This historic structure, which was, incidentally, the first Roman Catholic church in Idaho, was begun in 1847 by the Jesuit missionaries under the direction of Father Peter J. DeSmet. Church services began to be held in 1848, but the building was not finally completed until 1852. The Mission settlement remained for many years the outpost of Christian civilization in this wilderness, administering to the material as well as the spiritual welfare of the Coeur d'Alene Indians.[1]

After the Mullan Road had been completed, and a stream of travel and immigration (thin as the stream was) began to flow past the Mission, the Jesuit fathers removed their headquarters to a new mission at De Smet (in the present Benewah County). This removal took place in 1877. Thus, in 1880 the Cataldo Mission was almost deserted, only one priest, Father Joset, remaining in charge. Yet in its time the Old Mission had played the host to territorial governors, United States Army officers and troops, and hundreds of travelers. Even in 1880, the Mission was the outstanding landmark of the Coeur d'Alene valley.

Fragmentary records also tell of two inhabited locations on the South Fork of the Coeur d'Alene, namely the Miner's Cabin and Evolution. The Miners' Cabin embraced several small buildings on the Mullan Road about two miles east of the present city of Kellogg near the mouth of Montgomery Creek. Here Tom Irwin, an early prospector, had settled and claimed to have discovered and worked a quartz gold claim near the Cabin. This was in 1879. More of Tom Irwin later. Also a certain H. S. Blandford had made the Miners' Cabin his headquarters while at work stretching the military telegraph line from Fort Coeur d'Alene to Fort Missoula. Blandford was an operator and line repairer, and he made the Cabin a supply station for all necessary materials along the entire line from the Mission to the summit of the Bitter Root Mountains. At least the Miners' Cabin was known as a place for refreshment and recuperation by travelers of the Mullan Road.

Concerning Evolution not much is known definitely. A. J. Prichard, of whom there will be much more to say later, designated his winter camp of 1879–'80 as Evolution, located on the Mullan Road 23 miles east of the Mission, and indicated in a letter that he was three miles above the telegraph station at the Miners' Cabin. Evolution later became prominent as the terminus of one of the well-traveled trails from the South Fork over to the North Side, and therefore is more easily located definitely than the Miners' Cabin.

Thus, with these few exceptions, the Coeur d'Alene region in 1880 was uninhabited. It was a wild, rugged wilderness; heavily forested, inhospitable to man. A glance at the topographical contour map of the district or at the various photographs included in this study will reveal the rough and precipitous nature of the Coeur d'Alene Mountains, which are a lateral spur from the Bitter Root Range. However, a careful study of the lay of the land shows a certain uniformity of altitude and of slope in the ridges which has led geologists to describe the region as a great plateau, sharply cut by streams and by ancient glacial action. This plateau slopes from the high peaks in the east and southeast (Stevens, 6826 feet above sea level; Granite, 6810; Tiger, 6635; etc.) down to the comparatively low points of the river and lake levels (Enaville, 2160; Cataldo, 2143). Into this sloping plateau the two main forks of the Coeur d'Alene River and the hundreds of tributary creeks have cut deep valleys and gulches. Yet, on all these streams the descent is gradual; they are rushing mountain torrents, but they never pour over high falls. Likewise, there are no steep sheer precipices in the down-slope of the land, with the exception of a few glacial cirques on the north slopes of the high peaks.

Underneath the forested slopes lies the rocky backbone of the hills. Broad, deep layers of sedimentary rocks—slate, argyllite, quartzite—form the basic foundation structure of the Coeur d'Alene Mountains. Into this sedimentary formation has intruded a variety of igneous materials, of which we need concern ourselves only with the gold-bearing quartz ledges, and the rich veins of galena and carbonate lead-silver ores. Fragments of gold quartz, broken off from the hillside ledges and washed down by streams into gravelly banks and creek bottoms have formed rich placer grounds.[2]

The Coeur d'Alene region has a temperate climate with considerable rainfall in spring and autumn, with prevailing winds from the Pacific coast. Heavy snowfalls are common in the winters, but the winter temperatures are seldom severely cold. The summers are generally hot and dry, a condition which subjects the region to the serious hazard of forest fires.

Thus, by the tireless and deliberate processes of nature, the Coeur d'Alene region was prepared for a human drama on which we raise the curtain in 1880.

Robert Wayne Smith

1. The Coeur d'Alene Indians were a small tribe, numbering about 300. They were concentrated within a small area along the shores of Coeur d'Alene Lake and South of the lake. There was no Indian problem in the settlement of the Coeur d'Alene mining district.
2. The geological structure of the Coeur d'Alene district can be studied in detail from the United States Government publications; 1. Ransome and Calkins, *The Geology and Ore Deposits of the Coeur d'Alene District*. United States Geological Survey *Professional Paper* 62, (1908). 2. Umpleby and Jones, *Geology and Ore Deposits of Shoshone County, Idaho*. United States Geological Survey *Bulletin* 732, (1923).

CHAPTER III

The Discovery of Gold

Prospecting for gold was an old story in Idaho by 1880. Back in 1860, '61, and '62 the mines of the Clearwater had boomed; then came the rich discoveries of the Salmon River, the Boise basin, the Snake River, and finally Wood River. Every mountain stream in the territory which looked as if it might contain the golden sands was explored and tested. But Idaho was a large area, and the prospecting of the territory took a long time. The Coeur d'Alene district remained until comparatively late.

Yet even in early years there were hints of what was to occur later. Captain John Mullan commented on the quartz formations of the Coeur d'Alene and St. Joe valleys which he observed while engaged in building his road. Later, in a letter dated Washington, D. C., June 4, 1884, he wrote to Mr. A. F. Parker of Eagle City:

> "I am not at all surprised at the discovery of numerous rich gold deposits in your mountains, because both on the waters of the St. Joseph (St. Joe) and the Coeur d'Alene, when there many years ago, I frequently noticed vast masses of quartz strewing the ground, particularly on the St. Joseph River, and wide veins of Quartz projecting at numerous points along the line of my road along the Coeur d'Alene, all of which indicated the presence of gold. Nay, more: I now recall quite vividly the fact that one of my herders and hunters, a man by the name of Moise, a French Canadian, coming into camp one day with a handful of coarse gold, which he said he found on the waters of the north fork of the Coeur d'Alene while out hunting for our expedition. This gold was so pure, so heavy and so free from quartz or matrix rock and weighing several ounces that it attracted not only a great deal of attention, but some degree of surprise at the time. And when he said he found it his statement was scarcely credited by anyone then in our camp, because it was believed that he had traded for this gold with some parties passing up and down from the Frazier River mines. This was in 1858–'59.
>
> "***The members of my expedition were composed very largely of old miners from California, and having had more or less experience in noticing

the indication of mineral deposits, their universal verdict was that the entire country, from Coeur d'Alene Lake on toward and including the east slope of the Rocky Mountains, was one vast gold-bearing country, and I was always nervous as to the possible discovery of gold along the line of my road; and I am now frank to say that I did nothing to encourage its discovery at that time, for I feared that any rich discovery would lead to a general stampede of my men from my expedition, and thus destroy the probable consummation of my work during the time within which I desired to complete the same."[1]

We have already mentioned Tom Irwin, a pioneer prospector of the Miners' Cabin. Irwin had prospected all along the South Fork of the Coeur d'Alene River, keeping close to the Mullan Road, and in 1879 he discovered and worked a quartz gold claim near the Cabin. At least he claimed to have done so. Nothing definite can be said of Irwin's claim; it was never located with the county recorder. The South Fork has never yielded gold in paying quantities, yet a little gold has been found in the tributary gulches of the South Fork near the supposed location of the Miners' Cabin, particularly Elk Creek. It seems that by 1881 Irwin became discouraged at his lack of success. In that year he traded off his "grub" to W. Edson and Bob Fanning in exchange for a horse, on which beast he mounted and rode off for Colorado.[2]

There is still another record of early prospecting in the Coeur d'Alenes. Mr. John P. Vollmer of Lewiston asserts that he helped fit out a prospecting party to go into the Coeur d'Alenes in June, 1873, long before Tom Irwin's day. The party consisted of five men who worked the streams of both the Coeur d'Alene and St. Joe river systems as well as some territory in Montana. They claim to have found very rich gold quartz in the Coeur d'Alene valley, so rich in fact that Mr. Vollmer and his associates staked these prospectors for another expedition in 1874. On this second trip the party failed to locate their rich find of the previous year, after which failure no further efforts were made.[3]

Thus, gold was suspected, searched for, and perhaps found before 1880. But it remained for Andrew J. Prichard, more commonly known by his initials, A. J. Prichard, to make the great discovery which brought the

Coeur d'Alenes into the spotlight of national publicity. Prichard himself remains somewhat a man of mystery. His early history is obscure. His later life was uneventful; he patented a placer claim on Beaver Creek, died in 1902, and was buried in the cemetery at Murray. But for a brief moment he took the center of the stage and played a dramatic incident in the development of the Coeur d'Alene district.

A. J. Prichard was born in Connecticut, the exact place unknown, in 1830. His boyhood and youth were spent in Ohio where he married Louisa Poland early in the 1850's. Prichard and his young wife moved to Indiana, then to Minnesota, and finally to Missouri where they settled on a farm near Kidder, some fifty miles east of St. Joseph Missouri. His son Jesse was born at this place in 1856, and there were other children. When the Civil War broke out, A. J. Prichard served in the Missouri Home Guards for a year, and later he was a member of the 11th Missouri Cavalry, of the Union Army, in which service he rose to the rank of sergeant. After the war Prichard was widowed and traveled extensively in Kansas, Oklahoma, Colorado, and New Mexico, working at many jobs and occupations of which placer mining was only one.[4]

According to his own story, Prichard first came into the Coeur d'Alenes in the fall of 1878, traveling the Mullan Road from Montana, after having come from New Mexico and Colorado.[5] It was his habit to leave the road and travel the ridges, casting a speculative eye over the tributary gulches. In this manner he located in November, 1878, a quartz ledge which he called the Evolution lead. He tells of going on to Fort Coeur d'Alene and Hayden Lake where he worked some during the winter. He also went to Spokane Falls[6] in January, 1879, and undertook a timber contract for the firm of Cannon, Warner and Company. While he was at the Falls he met Tom Irwin for the first time, and together they discussed prospecting on the South Fork of the Coeur d'Alene. Prichard was kept busy on his logging work until the fall of '79 when he returned to Evolution and built himself a cabin where he spent the winter. In the meantime, Tom Irwin had also made his gold quartz strike near the Miners' Cabin, about two miles below Evolution.

In the early spring of 1880, Prichard began prospecting in a little gulch which flows into the South Fork at Evolution, called Prospect Gulch. He put in sluice boxes in March, and washed the gravel for as long a time as the water supply lasted. "I got some very nice specimens of coarse gold," he concludes, "which were, to my knowledge, the first placer gold on the Coeur d'Alene."

Prichard's own brief narrative ends there, almost as though he were deliberately concealing his subsequent actions. We must therefore follow his later movements through secondary and often conflicting sources. Mrs. Prichard, evidently repeating her husband's story, says that he remained in the South Fork country around Evolution all of the year 1880, prospecting and developing.[7] His second winter at Evolution followed.[8]

1881 was a year of great activity for A. J. Prichard. Mrs. Prichard's account and Adam Aulbach's version of the story and also James L. Onderdonk's *Resources of Idaho* all agree in the statement that Prichard crossed the divide north of Evolution and prospected the North Fork country for the first time.[9] According to Mrs. Prichard he went alone and spent all summer and fall.[10] His activities took him to Beaver Creek, Prichard Creek (named later by Prichard himself) and Eagle Creek. He prospected up Eagle Creek as far as Tributary Creek and up Prichard as far as the site of Murray. On this trip he discovered the rich placer grounds of Prichard Creek. From one pan of Prichard Creek dirt he washed $42.00! Not until the snows of winter made further work impossible did Prichard leave his new-found diggings and return to Evolution.

On other circumstance gives further color of certainty to 1881 as the year of the Prichard Creek discovery, to wit: when Prichard filed the notice of the Discovery placer claims, which was the earliest mining location made within the limits of the present Shoshone County, he gave the actual date of the discovery as July 12, 1881. This location was not filed, however, until March 25, 1883.[11]

The following June, 1882, saw Prichard lead a prospecting party to the North Side. Mrs. Prichard's account named M. Gelatt, Phil Markson, William Dempsey, and Fanning as her husband's associates. Adam Aulbach lists M. Gelatt, Phil Markson, and William Gerrard. We can be certain that Gelatt and Markson and perhaps Gerrard were in the party. After crossing the divide from Evolution, they descended Beaver Creek to the North Fork, cutting their way through a dense forest and undergrowth. Then up the North Fork to the mouth of Prichard Creek, and up the latter stream to the mouth of Eagle Creek, where they camped for the night.

Let us follow Adam Aulbach's spirited narrative at this point:

> "Two days later they made four miles up the creek (Prichard Creek) to a little flat about half a mile west of Murray. Here camp was made for a stationary base to prospect, and here is where the split occurred between Prichard and his associates, Markson and Gelatt. The real cause of the trouble was never made known by any of the men, except Markson stated that Prichard had become very domineering and desired each to do as he directed. Not desiring to quarrel with him, Markson said he left camp and proceeded a mile up stream, where he found a good prospect and concluded to locate and work it if the party remained. He returned at sundown to camp, just as Gelatt made his appearance. Gelatt had a small 'poke' full of coarse gold and it created quite an excitement in the little group. Prichard became particularly agitated, took the 'poke,' poured the gold into a pan and sized the weight. Prichard then walked a few steps to the stream and declared, 'I christen this water "Prichard Creek." Gelatt detailed the place of his discovery as about half a mile distant, across the creek, and rather high on the rim, with scarcely any surface covering. The claim subsequently made local history. On the spot where the camp was located Prichard drove a stake and called the ground 'Discovery claim.' Markson said he was only a quiet spectator. Prichard became excited, took the 'poke' full of gold and informed his companions that he would start for Rathdrum early in the morning to buy provisions and return as soon as possible. He bought the provisions and sent them forward to Evolution, and had them packed to Beaver Creek. From there men brought them to Prichard Creek."[12]

Thus was Prichard's discovery of the previous year verified and the location made for filing of notice with the county officials at Pierce City. Gelatt also located a claim on the rich grounds and so did Markson.

Robert Wayne Smith

1. This letter of Captain Mullan is contained in Onderdonk's *Idaho,* pages 33 and 34.
2. The Irwin story is told by Adam Aulbach in his newspaper article, "Coeur d'Alene, Idaho's Famous District Capable of Producing Ten Millions a Year." *Portland Oregonian*, 1897.
3. Vollmer's letter to the editors of the *Illustrated History of North Idaho.* (Henderson, Shiach, Averill) 1903. pp. 985–986.
4. These fragmentary facts, never previously published, were gathered in the course of an interview with Jesse A. Prichard at Delta, July 3, 1932. Jesse Prichard is himself an old man (72 years of age) in feeble health. He did not come to the Coeur d'Alenes until 1901, only a year before his father's death.

   In this connection, see also Appendix I, an obituary of A. J. Prichard, published by the Wallace Press, October 11, 1902.
5. Prichard's *Narrative* in Appendix II. Also, *History of North Idaho*, p. 985.
6. Spokane Falls, Washington Territory, now the city of Spokane, Washington. Its population in 1883 was 1,000; in 1885, 3,500.
7. The second Mrs. Prichard, of course, who was formerly Mrs. Katheryn Whistler, a restaurant and hotel proprietress of Eagle City.
8. Mrs. Prichard's remarks quoted indirectly, the *History of North Idaho*, pp. 986–987.
9. *Idaho. Facts and Statistics concerning its Mining, Farming, Lumbering, and other Resources and Industries.* James L. Onderdonk, Territorial Controller. San Francisco, 1885.
10. Aulbach asserts that Edson and Fanning, who acquired Tom Irwin's property, accompanied Prichard; that they prospected Beaver Creek, but found nothing; after which they turned over their outfits to Prichard and left the country.
11. Newspaper interview by Harry A. Rogers, auditor and recorder of Shoshone County, Idaho, in Wallace Miner, 1925.
12. Aulbach: "Authentic story of the Discovery of Gold in the Coeur d'Alenes." *Wallace Miner*, February 25, 1932. Aulbach obviously considers this incident the original discovery, and Gelatt the real discoverer; but the events of 1881 already chronicled seem more convincing for that date.

CHAPTER IV

The Rush of 1883–1884

A. J. Prichard succeeded well in keeping secret his discovery of 1881. But when he appeared in Rathdrum[1] and Spokane Falls in the summer and fall or 1882 with a well-filled "poke" of gold dust, the story began to leak out. William T. Stoll in his interesting but rather inaccurate book, *Silver Strike*, tells of seeing Prichard in a bar-room in Spokane in August, 1883, carrying four pounds of gold dust.[2] Prichard had little to say; but the four pounds of dust fairly shouted an invitation to the curious throng.

Back at Evolution in the winter of 1882–'83, Prichard toiled through the days and nights writing letters telling of his discovery. How many he wrote we cannot say, but he reached a considerable audience directly or indirectly. Two samples of Prichard's correspondence are to be found in the Appendix, numbers III and IV. They bear dates close to each other and their content and tone are similar. The letter of Appendix III is undoubtedly genuine, the original, in Prichard's handwriting, being in the possession of Mr. Jerome J. Day of Wallace. The letter of Appendix IV has not so firm a ground of authenticity. It was published by the *Spokane Review* in 1888. In both letters, Prichard described his discovery and invited certain friends to join him in the new gold fields while choice claims could still be secured. The recipients of these letters were all residents of the Gallatin valley in Montana, and members of a certain Liberal League to which Prichard also belonged. This Liberal League was a small body of men who termed themselves "free-thinkers," and agitated, though without much success, doctrines embracing atheism and discontent with the existing social order. The Liberal League published a paper called *The Truthseeker* in whose pages at least one of the Prichard letters received publication.

Prichard urged extreme caution and secrecy upon his correspondents, but apparently his admonitions went unheeded. The rumor of gold mines in the Coeur d'Alene region began to spread through Montana and the Black Hills as it had already begun to circulate in Washington and Idaho Territories. Accordingly, in the spring of 1883 there appeared an eager crowd of gold-seekers, most of them totally unequipped for mining.[3] They found Prichard at Evolution, and demanded that he lead them to the mines. This he refused to do, arguing that the trip was too difficult at that season of the year, and furthermore that it would be useless because of lingering snow and high water. The mob coaxed and bullied, and finally threatened to hang Prichard if he didn't lead them to the gold. He yielded and the party swarmed over to the North Side. Prichard conducted the group to the discovery sites on Prichard Creek. His predictions were realized; the water on the creek ran so high that bed rock could not be reached. Most of the gold-seekers quit the country in disgust. Others settled down to wait for low water, establishing a camp at the mouth of Eagle Creek. As soon as conditions were favorable they began prospecting, and some of them made rich finds.

As the summer of 1883 wore on, more and more people came into the Coeur d'Alenes to prospect and to locate claims. Prichard had filed his Discovery claim on March 25, and he proceeded to make additional claims for his absent friends, using a power of attorney to do this. In fact, the Discovery location of March 25 was in reality a group of four claims of twenty acres each—80 acres in all—one of them for his son Jesse A. Prichard; another for Mrs. C. A. Schultz of Michigan (this the Pacific claim, later famous as the Murray claim); still another for Mrs. Mary H. Lane of Illinois (the famous widow claim); and the fourth for Willard O. Endicott, a son of Prichard's old lieutenant of Civil War days. In each of these claims, Prichard was to have half interest for working and management.

But the rush of the spring and summer of 1883 was a small affair compared with the stampede of the ensuing fall, winter, and spring. Coeur d'Alene and Eagle City were magic words throughout the northwest in 1883 and '84. The chief responsibility for the great rush lies with the

Northern Pacific Railroad which had just completed its transcontinental line in 1883. The mainline of the railroad passes down the Clark Fork valley in western Montana, along the north shore of Pend' Oreille Lake in northern Idaho, and westward to Spokane. It thus sweeps in a broad arc around three sides of the Coeur d'Alene district. Mr. H. C. Davis, assistant general passenger agent, as soon as the Coeur d'Alene discoveries had been confirmed, in the summer of 1883, conceived the idea of stimulating passenger traffic by advertising the new mines. His method of advertising was to print a sensational circular and scatter the copies broadcast over the country.

The Northern Pacific gold circular (see Appendix V) should be read in full. The Coeur d'Alene mines were described as a second California, a new El Dorado, in which there was a fortune waiting for everybody. The circular's glowing phrases were studded with superlatives and with flashy quotations of fabulous values. To call the circular an exaggeration is to speak mildly. In reality it was a *magnum opus* of the imagination based upon a few known facts. However, it spoke a welcome language to its readers. The name of Coeur d'Alene became almost a household word, and restless adventurers all over the land were kindled to a feverish excitement to rush in and share the abundant wealth.

From October, 1883, to April, 1884, literally thousands of fortune hunters rushed into the Coeur d'Alenes. Estimates vary from 2,000 to 10,000, with a reasonable figure somewhere about half way, 6,000. Eagle City was the Mecca of this pilgrimage, and the devout came in every conceivable manner and route. The railroad circular was issued so late in the year that those who read it arrived too late to test the placer grounds. A heavy blanket of snow covered all the country, lying ten to fifteen feet deep in the higher mountains. There were no roads; trails were poorly marked; the weather was bitter cold for long stretches; accommodations at Eagle City and the other camps were very meager. Still they came, trudging afoot over snowy trails and carrying packs; or leading a pack horse or mule; or dragging a toboggan. There were veteran miners from Montana and the

Black Hills, from Oregon and California. In fact practically every famous mining camp of the west had its representatives in the Coeur d'Alene that winter. In addition, there were the usual number of tenderfeet and novices attracted by the lure of easy money.

The question arises, how did the gold seekers of 1883 and '84 reach the mines, i.e., what were the routes? There were several. The first route was the one already mentioned as having been made by Prichard and other prospectors. It left the Mullan Road at Evolution and crossed the ridge to the north into Beaver Creek; then wound down Beaver Creek to the North Fork and up the river along the south bank to Prichard Creek, and up Prichard Creek to Eagle City at the mouth of Eagle Creek. Travelers from the west had to take wagons or ride horseback from Spokane or Rathdrum on the railroad line to Fort Coeur d'Alene, thence on to the Mission, and up to Evolution before they could begin the final trek over to the North Side.

This is the route taken by William T. Stoll who followed it in November, 1883, as told in his recent book, "Silver Strike." The party in which he traveled took a long day to go from Fort Coeur d'Alene to the Mission. The next day they walked to Evolution, which Stoll described as "a station where one saloon herded a score of tents." A furious snow storm held the party at Evolution for three days. The fourth day they struggled through snow which had drifted six to ten feet deep; a bitter cold wind howling over the snow, and nearly reached Eagle, being forced to camp in the woods when darkness overtook them while they were trying to make a short-cut over the ridge. The next morning they moved on into Eagle City.

From the west there was another possible route to the mines. Leaving the Mullan Road at Kingston, one could travel up the North Fork all the way to Prichard Creek and thence to Eagle over the same route as the Evolution trail. The North Fork route, while it eliminated the high ridge between Evolution and Beaver Creek, was long and poorly marked, and hence was not popular with the immigrants until the spring of 1884 when it became practicable for small boats (bateaux).

Four stations on the Northern Pacific in Montana became termini for trails across the Bitter Root divide into the North Side of the Coeur d'Alenes, namely Thompson Falls, Belknap, Heron and Trout Creek. Thompson Falls was not laid out until the beginning of 1884, but immediately thereafter it was recognized as the most practicable point of entry from the railroad on the east. By the middle of March, 1884, a sleigh road was under construction, and in the summer a good trail was put through Thompson Pass into the upper valley of Prichard Creek.

The trails from Belknap and Heron enjoyed brief popularity, but the most traveled of all the Montana trails that first winter was the one from Trout Creek. The route lay up Trout Creek valley on the Montana side, over the Bitter Root divide at an elevation of 6,000 feet above sea level, and down the length of Eagle Creek to Eagle City, a distance of 35 miles. The trail had been made by Joseph Hamilton of Missoula and P. Tone, former deputy sheriff of Helena.[4] Considering both the season of the year and the altitude, the Trout Creek trail was essentially a snow trail, and it was the favorite route for freighting supplies on toboggans. The trail carried an immense traffic. One traveler counted over 100 toboggans loaded with freight on the trail in one day. People without heavy loads might make the trip in one day. Two to four days were required to drag toboggans over. 125 pounds were considered enough for a man of ordinary strength to carry. For people who took mote than one day to cross, there were stations (tents) en route at which meals sold for 75¢ and lodgings went for one dollar. The latter charge was modified to the effect that people with their own blankets could spread them on the floor without cost. There was a toll charged on the Trout Creek trail, 50¢ from parties going toward Eagle, and 25¢ from those going toward Trout Creek. In addition to the great number of men traveling the trail, there were "many women, a few children, and two or three lame persons" who crossed, according to the account published March 22, 1884.

A brief but interesting item appeared in the *Coeur d'Alene Nugget* for March 15, '84, as follows: "Charles Riar, the well-known Black Hills pedestrian, is in town. He has been in nine walking matches, eight of which he won, the most important contest being in Chicago, Riar wishes to wager $2500.00 that he can walk from Trout Creek to Eagle over the trail, in ten hours, and carry one hundred pounds of flour."

Later in the spring, the Trout Creek trail became impassible with melting snow, and soon after lost its preeminence to the Thompson Falls route.

1. Rathdrum, Idaho, a small town on the main line of the Northern Pacific Railroad, about the closest railroad point on the western side of the Coeur d'Alenes. In the 1880's Rathdrum was county seat of Kootenai County, Idaho. In 1890 its population was 250.
2. *Silver Strike*. The True Story of Silver Mining in the Coeur d'Alenes. As told by William T. Stoll to H. W. Whicker 1932. A book of reminiscences apparently dictated by Stoll and published without a critical reading. Full of inaccuracies.
3. All the records tell of this premature rush, but the date of its occurrence varies as does the date of the Prichard Creek discovery. Aulbach puts it in 1882, before the discovery. Onderdonk and the *History of North Idaho* say 1883, the latter giving an account by one of the mob.
4. The details of the Trout Creek trail story are all gleaned from the leading article of the *Coeur d'alene Nugget*, Vol. I, No. 2, March 22, 1884, Eagle City.

The Grave of A. J. Prichard, at Murray, Idaho.

CHAPTER V

Eagle City

The first, and for a season the largest mining camp in the Coeur d'Alenes was Eagle City. Founded in November, 1883, Eagle became the metropolis of the mines, harboring between 2,000 and 4,000 people by the following spring. The townsite is one of the most beautiful imaginable, lying at the confluence of Eagle and Prichard Creeks on a wide flat area surrounded by heavily forested hills. Not a vestige of the old town remains today, and no photographs are extant. Yet it must have presented a picturesque appearance in its one big boom winter. The two principal streets made a right angle in the center of town. Eagle Street extended up Eagle Creek to the north, and Prichard Street ran east up the valley of Prichard Creek. The settled area extended up Eagle Creek as far as the forks of the creek, about a mile and a half or two miles. Town lots were laid out and sold at high prices. Cabins, tents and business houses rose as quickly as they could be constructed. There was a general rush of business of all kinds common to a stampede mining camp. Its rise was meteoric, and its decline almost as rapid.

The best pictures we can get of Eagle City are those drawn in words by the pioneer newspapers. The first newspaper was the *Coeur d'Alene Nugget*, published weekly by C. F. McGlashan and W. F. Edwards. Their first number appeared Saturday, March 15, 1884. A photostatic copy of the front page of this first number of the *Nugget* appears here in this study. At least nine numbers were published, the last known issue being dated May 7. The first four numbers were printed in Spokane on the press of the *Spokane Falls Review*, but by April 9, the *Nugget's* own printing outfit had been set up and had published number five, the first newspaper actually printed in the Coeur d'Alenes. Before the end of April, 1884, another

The Site of Eagle City, Idaho.

pioneer newspaper was in circulation, *The Eagle,* of which A. F. Parker was editor. The *Eagle* is known to have been in existence in June. Probably both the *Nugget* and the *Eagle* continued publication through the summer of 1884, since the first term of the district court in Eagle City furnished vital news in July and August. Long before mid-summer, however, Eagle City had been almost abandoned, and the center of mining interest had shifted to Murray.

In order to give the reader as comprehensive a view as possible of actual conditions in Eagle City, I can do no better than to quote, judiciously, I hope, from the columns of the *Nugget* and the *Eagle*.

The general picture may not be a distinctive one, for Eagle City was probably typical of mining boom towns all over the west, but in its detail Eagle City presents some amusing and interesting features.

He is a general view of the town, from the *Nugget* of March 15:

Eagle City (editorial)

The metropolis of the Coeur d'Alene needs no word of praise. Beautifully located, peopled with men of energy and enterprise, and offering superior inducements to capital, she has burst into a city in a day. To number the business houses this week gives no idea of the number that will be established next week. People who were here a fortnight ago was hardly recognize the place to-day. Houses and tents appear like magic where yesterday were only snowdrifts or tamaracks. New signs, painted on canvas, peep up in the most unexpected places and give legends concerning the most unexpected occupations. "Tin Shop," "Jeweler," "Hardware," "Dance House," "Theater," and the like, confront one so suddenly as to excite an exclamation of astonishment. Everything is new, but all denotes enterprise and permanence. Some very fine houses are being erected. Buildings that will be no disgrace when brick and lumber have superceded logs are being put up in all quarters of the town. The advent of a sawmill would greatly accelerate the growth of the city, but if sawmills are delayed people will not wait for their coming. A city full of people has arisen from the snow as if by enchantment, and the strangest part of all is that her present growth is not even an intimation of that to come.

A parallel account from an April issue of the *Eagle* gives more details:

At this stage of its growth and development, Eagle presents the appearance of a "hard" place. Its buildings are located on lots from which snow to the depth of four feet has been excavated and dumped into the street, so that between the buildings and the streets there is a solid wall of snow. The buildings are composed of logs and shakes. Great tents with gaudily painted signs loom up in endless variety before the spectator,while from within come the sound of revelry and strains of music, the click of chips and the metallic chink of hard cash as it passes over the bar. The streets and places of public resort are thronged night and day with miners and prospectors waiting for the snow to disappear so they can get off into the mountains on their annual prospecting tour. Then there are the packers and transient population, a tough looking crowd, but very orderly, who stand around enjoying the luxury of a sun bath. On the corners are knots of men talking mines and mining, and criticizing ore specimens that pass from hand to hand. The stores are thronged discussing the locality and merits of the last new thing in rich strikes; the centers are strewn with specimens of quartz from the different districts; while on the street there is an ever changing panorama of countenances as the crowd diverts itself. The report of a pistol shot will bring a hundred men to their feet for an instant, and the saloons will disgorge twice as many more in the same moment, all on the alert to catch a sensation, which has never occurred yet. Then there are the arrival of pack trains and toboggans, in a more or less advanced stage of dilapidation from hard usage on the different trails. On all hands carpenters and mechanics are busy erecting shanties for temporary occupancy, but talk with whom you will you find that the utmost confidence is expressed in the outcome of the camp; that we have the mineral here to make the biggest camp in America and that the business men are conservative enough not to be guilty of exaggerating the mineral wealth of our placer fields and quartz ledges in order to encourage a wild stampede. On all hands there is a general disposition to tell the truth, to let the camp sustain itself and to await developments on the more promising quartz prospects before booming the camp. The business men of Eagle include representatives of every state in the union, and it is a promising sign that they are purchasing town property and mining ground and are all preparing to erect larger and more commodious places to business as soon as the material can be procured. This is about the status of Eagle to-day. Every branch of business is well represented. We have two banks, several stores of general merchandise and more saloons and lodging houses than you can shake a stick at.

Thousands of dollars worth of real estate transfers are made and recorded daily, and thousands more are being invested in substantial buildings and other business enterprises. No lawlessness exists.

An extended quotation from the March 15 *Nugget* describes the populace, with a special encomium for the mining fraternity:

The Motley Throng to Be Seen on Our Streets.

Among the three thousand people who have thus early found their way to the Coeur d'Alene mines, a large proportion are business men. They are men of means, of energy, of intelligence, who come here to obtain the first opportunities of investment which are so desirable in all new camps. They are shrewd, thoroughly informed men, who eagerly compete with each other for the numerous business prizes which are to be drawn in this great mining excitement lottery. There are hundreds of laborers here. Hewers of wood, and men who are content to earn honest wages for honest work. They are stalwart, sinewy fellows, who toil early and late, in snow and storm or sunshine. These are they who are building cities in these desolate mountain wilds. There are gamblers here. Of that rest assured. There are hundreds of them. Gamblers of the genuine Bret Harte style. Men of large hearts, of generous impulses, men to whom, of all that are in camp, you would go first to ask a dollar with which to buy a meal. Men who would never refuse you so long as they had the dollar. Gamblers there are, too, who are not of this stamp. Men who are constantly begging stakes of the better class with which to fleece victims. And the victims are here. Lots of them. Some are workmen, and lose their wages every week. Some bring a little money with them to the Coeur d'Alene, and return home broke. The "girls" are here too, and more of them some every day or two. Some of them dress stylishly and wear silks and diamonds. Some dress coarsely and slovenly. Quite a number wear men's clothes, and walk the streets in garments which would excite the envy of Susan B. Anthony. Boys are here, mere youths not yet out of their teens. It is a terrible school for lads, yet they have come with the rest. There are old '49-ers here. Men whose hair and beard have been silvered in the tireless, persistent search for gold which will never end till life's close. They are all prospectors, and you find them living a hermit's life in little cabins up lonely gulches. And the miners are here. The true, grand-souled old-time miners, with their blue shirts and open-palmed hands. Their cabin, their fare, their blankets will be shared with the stranger just as it used to be in the golden days on the American and Yuba, just as it always has been where there were miners. If any class should be monarch among men, it is these same miners. They laugh at hardships and privations, they never think of shirking work or of doing a dishonorable act, and when the dust is panned out, it goes as easily as water. Bye and bye many a rough, hard character will come, many a bully, cutthroat, scoundrel, but so long as the miners are in the majority, the camp will be all right. The miners are the men upon whom depends the fate of the country, the men, of all others, who will rule the camp. Ah! yes, and another has come, the Angel of Death. Neither the snow, not the rugged mountain, nor the silent forests, nor the deep dreary canyons could hide from him this new born camp. He has come, and his pinions have borne some of our number beyond the worry and weariness of this struggle for gold. Under the hillside snow lie four of the boys, who came eager and thirsting, little dreaming that their bodies evermore would rest among the gold dust they came to find. Yes, the Angel of Death is here, and scores and hundreds will acknowledge his presence ere the year closes, and next winter's snow will find them sleeping under the snow, among the golden sands.

A few brief notes on the personnel of the camp:

It is nothing unusual for two or three hundred men to come into camp in a day, over the different trails. No attention whatever is paid to the new arrivals as they file through the streets.

S. Haptonstall and E. A. Boblett, two of the original locators on Prichard Creek, have returned to the mines, and are looking after their property. Mrs. Boblett came to the mines with her husband in May, 1883, walking and leading a pack horse from the Mission. She was the first white woman in the mines, and remained until November.

Mrs. M. Chandler of Osborne was the only white woman who wintered in the Coeur d'Alene. She and her children, two in number, came over the Evolution trail in October, and located about two miles above Eagle. Until the trails became impassable, the boys of Eagle used to send their Clothes up to Mrs. Chandler to be washed.

Food, shelter, and other necessities of the mining community provide interesting items, as follows:

Two prospectors living a few miles down the river, have killed thirty-five deer this winter.

The Pioneer Restaurant is so over crowded that from fifty to one hundred customers are turned away every day. As meals are $1.00 each, it can readily be understood that the above indicates a rushing business.

The trees suitable for building purposes in the immediate vicinity of Eagle have been cut down, and contractors complain that logs have to be hauled from a considerable distance, or chuted down the mountain sides from a great height.

It is estimated that one thousand buildings are in process of construction in the mining district. Every man who will work finds ready employment at from $5.00 to $8.00 per day. There is any abundance of small rigid timber suitable for log houses, and the sound of axes and falling trees can be heard in every gulch, and from every hill side. Cabins are costing from $250.00 to $2000.00, every log delivered on the ground being worth $5.00. Lumber whipsawed is worth $250.00 a thousand and finds ready sale. Shakes are $3.00 per hundred the cedar being the best adapted to their manufacture.

Frank Reed's house 20 x 35 feet in the clear, 2 stories high, cost $1200.00 and rents for $300.00 per month, on a six months lease. Another 25 x 35 house, 2 stories high, with double glass doors, rents for $400.00 per month. The money was paid in advance before the foundations were laid.

One enterprising individual brought a box of apples over the Trout Creek trail and sold them three for a half dollar!

Men in cabins can batch at from $4.00 to $7.00 per week; some men can live on 2 bits a day. These are exceptions. It costs the average man $5.00 a day to live in Eagle. However, work is plentiful and wages are good.

Almost every cabin has windows and a few houses have double glass doors.

Warmth and light were particularly expensive items in Eagle City:

Stoves and windows come very high. Sibley stoves (a conical piece of sheet iron with a door and a hole for a chimney) cost $30.00 to $40.00; sheet iron box stoves, $20.00 to $80.00. Ordinary cooking stoves bring $75.00 to $150.00, and a range suitable for a restaurant is unobtainable, but it would be worth $250.00 to $400.00.

Window sash costs $1.00 per light, and window glass $1.00 per pane. These are the lowest prices.

Further quotation of prices should be of interest:

Eagle City Prices

(quotations from Sam'l H. Hayes, local merchant.)
flour per bbl. $60.00
flour per sack $15.00
coffee per lb 50¢
sugar per lb 50¢
beans per lb 50¢
candles per lb 50¢
dried fruits per lb 50¢
tea per lb $1.25
bacon per lb 40¢
tobacco per lb $1.00
potatoes per lb 25¢
lard per lb 40¢
soap per lb 25¢
assorted jams per lb 75¢
yeast powder per lb 75¢
spiced bacon per lb 50¢
ham per lb 50¢
soda per lb 25¢
canned fruits per 2½ lb $1.00
cheese per lb 25¢

salt per lb 25¢
crackers per lb 30¢
butter per lb 75¢
corned beef 2lb $1.00
syrup per gal. $4.50
coal oil per can $7.50
Jamaica ginger 3 bottles $1.00
cigarettes 3 papers 50¢
cigars per M from $40.00 to $150.00
liquors per gal $8.00
saws per foot $1.00
whip saws from $12.00 to $16.00
stoves from $40.00 to $150.00
underwear, flannel $5.00
brooms each $1.25
shovels each $2.50
picks each $2.50 to $5.00
axes each $2.50
buckets each $1.75
ax helves each 75¢
canvas per yard 30¢
house lining per yard 15¢
overshirts $2.50 to $5.50
boots per pair $7.00 to $12.00
boots, gum, per pair, $8.00

The interest in mining naturally continued at fever heat in Eagle. Here are some evidences:

> The largest nugget picked up on the Wyant claim weighed $140.00. Several have been found worth $20.00 to $54.00, and nuggets weighing $2.00 to $5.00 are frequent.
>
> Nuggets the size of bullets have been brought down from Raven.
>
> The boys who have charge to the Mother Lode have a unique prop to keep their door open. It is a mammoth piece of quartz, one side of which is full of glistening gold. The specimen is called the "snowflake."
>
> Nearly one thousand claims have been put on the records by Col. Frank

Points, recorder.

Wm. M. Sutherland, owner of the "Sunset" ledge on the headwaters of Beaver, has a ledge from 5½ to 7 feet wide in sight, and it can be traced on the surface for a distance of 1200 feet. A number of assays have been made, with the results varying from $25.00 to $2,000.00. The rock yields about 78 per cent lead, and is the first authentic discovery of galena in the new mines.

An interesting comment or two on the monetary situation:

As high as fifteen per cent premium is paid for silver, and a cart load would find sale at ten per cent. Mr. Denny has sent out a man with $500.00 to bring back silver, and other business men will soon be compelled to follow his example. Saloons and business houses lose on an average from $20.00 to $30.00 per day by not having change.

Silver and one dollar bills are so scarce in the mines that it is extremely difficult to get change for a five. Change is decidedly at a premium and parties coming here would do well to bring in a supply. Merchants and business men are very accommodating to each other, but all are constantly cramped for change in conducting their transactions.

Other businesses beside mining were flourishing, as the following quotations indicate:

Hood and Company, at enormous expense, with all imaginable difficulty, and in the face of all conceivable obstacles, have established the first sawmill at Eagle. They have a steam outfit with a capacity for 14,000 feet in 24 hours. The outfit was brought on sleighs through the Fourth of July canyon, and by boats up the North Fork. The mill is being located on Eagle Creek just below Eagle City and will be ready to furnish lumber by April 10.

Roram and Brown's sawmill is coming up the North Fork in boats.

Captain Friday opened his Chicago Laundry.

Twenty new business houses opened in the week of March 22. All are doing rush business.

A finely inlaid door ornaments the entrance to Ainsworth, Hawkins & Co's Bank. It is decidedly the handsomest door in the mines. The bank office is neatly papered, and the floor is covered with matting. It is as cosy and snug a room as one could wish to find anywhere.

The *Nugget's* comments on its own enterprise:

> Our material began to arrive Thursday, April 3rd, ahead of any other outfit. All our material was here Friday, April 4th, and there was not a pound of any other outfit in town. Our press came Saturday and was the first landed in Eagle. And now we present you the first paper.
>
> It has not been easy to issue this number. We can't stop across the street to a type foundry and get this or that little thing that was omitted in the outfit. A stovepipe makes a pretty good roller mould, cigar boxes furnish excellent reglets, nonpareil leaders are a passable substitute for advertising dashes, the top of a stove answers pretty well for an imposing stone, and the bottom of a dripping pan is a capital distributing disc.

Throughout the papers we find a preoccupation with problems of transportation and communication:

> Seventy-five chinamen have been at work on the Belknap wagon road.
>
> Judge P. Week's lodging house outfit came in last week, on the backs of Flathead Indians. Two of the Indians carried 125 pounds each, and another carried 143 pounds. They came over the Thompson Falls trail, and were three days making the trip.
>
> Noble's outfit of 3,000 pounds of wines, liquors and cigars came in over the Trout Creek Trail. Ten men drew 200 pounds apiece with no breakage.
>
> Ten dogs have been put to work hauling freight over the Trout Creek route, each dog hauling 30 to 40 pounds. There are no sleds, but two shafts are harnessed to the dog's shoulders with the load fastened behind. The dogs have provided very dependable transportation.
>
> Letters cost 50¢ apiece, and take up to two weeks to come from Spokane Falls. There is a weekly mail service from Fort Coeur d'Alene. Business houses pay $30.00 a month for mail service, others must pay 50¢ for each letter.

The postal difficulties were soon corrected. In April a post office was established at Eagle, A. F. Parker being appointed post master, and mail service three times a week was inaugurated between Coeur d'Alene and Eagle.

Robert Wayne Smith

Launching the Coeur d'Alene

The Coeur d'Alene city correspondent of the *Oregonian* under date of March 27th says: "The launching of the beautiful stern-wheel steamboat Coeur d'Alene is a great event in the history of Northern Idaho. This elegant river boat, which has engaged the labor of a large force of mechanics all winter has at last received the finishing stroke of the hammer in the hands of her skillful engineer, Henry Pope. She was launched this morning within 100 feet of her wharf and warehouse, and directly in front of the Exchange and Tubbs hotel, and within rifle shot of the military headquarters of the post. Crowds of anxious spectators were on hand long before the hour designated to enjoy the novel scene. A number of invited guests of both sexes availed themselves of the pleasure of going on board, and under the supervision of her experienced Captain J. B. Sanborn, and the able management of her foreman Mr. James Hanrahan, when the signal was given she glided down the ways into her future element as gracefully as a ball moves on the rail of a ten-pin alley. This boat is 146 feet in length. Her engines are 240 horse power, and she is of 130 tons burden. She has ample room on her deck to carry comfortably 150 passengers. Her cost will approximate $23,000.00 She will make her trial trip in a few days up the lake and river, returning in the afternoon. Her first freight and passenger trip will be made on or about the first of April, and she is a welcome boon to the many who for days and weeks remained at Coeur d'Alene anxiously awaiting her final completion. She will make her landing within fourteen miles of Eagle City, as the driftwood and other obstructions in the river are cleared away."

The Trial Trip

One of the best signs for the camp is found in the statement of Col. L. S. Howlett who arrived here Sunday night. He says that Sanborn's steamer, the Coeur d'alene, came up as far as Kingston, the new town at Long prairie, and added that she handled beautifully. The Captain made an experimental trip, breaking through the ice of the lake on Tuesday and Wednesday last. In good water she will run up nearer the camp, and in any event will greatly facilitate the transportation for passengers and freight from all points. The "Coeur d'Alene" is described by Col. Howlett as built "for Keeps," and furnished throughout in the best style. The owners have looked the ground over, and it is clear that they fully believe in the camp. They will find that the camp reciprocates.

This steamer began operating between Coeur d'Alene and the Mission early in April, 1884, and for many years provided the best and most comfortable transportation into and from the Coeur d'Alenes.

The liquor business, and other recreations and sports of the miners fill considerable space in the *Nugget*. The saloons bought the largest advertising space in the papers, and they were quite obviously the centers of community life during the winter.

Coy and Hess opened the "daisy" at the corner of Placer and Eagle streets February 26th, and a thousand men were in the house during the evening. The Daisy is a tent-covered building thirty by fifty feet, and is well furnished with bar, tables, etc.

Coy and Hess Bar receipts for their first 15 days of business amounted to $4,680.00. Their doors were never closed. They employ seven men in and about the saloon alone.

Messrs. Coy and Hess have exhibited great enterprise in introducing the first bottled beer. The call was so great that the cases first received were soon emptied. They have made arrangements however, for an ample supply.

A dance house is in full blast on Eagle street, a large number of "girls" having arrived.

Wanted immediately, ten first-class variety ladies, good dressers, and of handsome figure on the stage. Address J. Mc Daniels, Eagle City.

The faro dealers of Eagle are having a serious time. A number of them were broken by the boys last week, and all of them lost heavily.

The Comstock, a mamouth saloon, measuring 30 feet by 80 feet, opened at Eagle last week. 1500 guests were present at the opening. The best brands were served up by the most expert mixologists. A fine band is in attendance. There are twelve tables for cards.

$10,000.00 a day changes hands at the various games of chance.

There may be a famine in the mines, flour and beans may become scarce articles, but thank goodness the supply of whisky is ample for all contingencies.

In this connection, Adam Aulbach has remarked, "The amount of spirits freighted into the camps at rates ranging from 3¢ to 20¢ a pound exceeded the shipments of food."

Eagle City Theatre

J. Mc. Daniels, who is known world-wide as the "Barnum of the West," has established a theatre in Eagle City, and the opening night is placed at March 22d. He is putting up a theatre 30 x 120 feet, and will give the gold mines a first-class entertainment from the outset. It is his intention to give a performance every night with an entire change of programme each evening. The best talent has already been engaged, and the Eagle City Theatre will be the chief attraction of the town. Mr. Mc Daniels has been a theatrical manager, for twenty-five years, and his thousands of old friends from Quincy, Illinois, St. Joseph, Missouri, Omaha and Julesburg in Nebraska, Cheyenne, Wyoming, Deadwood, Dakota, Leadville, Colorado, Kansas City, Missouri, El Paso, Texas, and the numerous other cities in which he has been the proprietor of large and popular theatres will be glad to know that so genial and accomplished a showman is to furnish entertainment to the miners of out new El Dorado. He spent twelve years in Cheyenne, two years in Leadville, and was identified with other localities for considerable periods of time. His credentials include the most flattering and substantial proofs of his worth, energy and success, and show that he never left a place without causing sincere regret by his departure. He is putting forth his best efforts to give our citizens the choicest programme he has ever offered to the public. And his long and varied experience peculiarly recommend him as exactly the right man in the right place. Eagle City is to be congratulated on receiving such an acquisition as Mr. Mc Daniels and this troupe, and a new era in the mining camps will date from the opening of his theatre. Mr. Sam'l

H. Hayes and Tay Buzard are the financial backers of Mr. Mc Daniels and will furnish the sinews of war in unlimited quantities enabling to do his level best from the opening night.

The Wrestling Match

The new addition to the Daisy was christened Sunday night by a wrestling match between E. E. Coy and F. M. Rogers for two hundred dollars and the gate money. At the appointed hour the large tent was well filled with spectators. John Peyton and Alex Canutt were selected as judges, and Jack Dowd, the well-known sporting authority, was appointed referee. Prior to the wrestling match there were two lively sprints with boxing gloves. Noell Gallanon and Joe Campbell fought four rounds, and Nat Butts and Mike McMahon three. The sparring was lively and interesting, and the latter contest was especially spirited for amateurs. The wrestling match was side hold, best two in three. Mr. Rogers was the heavier man, and in the best condition, Mr. Coy having been seriously indisposed during the past few days. The struggle between the two was brief, but very exciting, and proved that each of the contestants is an adept at the science. Mr. Rogers was victorious in two successive rounds, and at the close issued a challenge to any man in the world of his weight, for any sum of money, side hold wrestling. The entertainment was a spicy break in the usual monotony of camp life.

From all the records available, Eagle City was a peaceful and orderly camp. Violence was rare.

The throngs of men to be seen in every saloon and business house during the evening astonish all who are not accustomed to mining excitements. Standing room can hardly be obtained in any of the popular resorts. Gambling in all its forms is carried on as recklessly as can well be imagined. Hundreds and thousands of dollars change hands every hour. And yet there is less quarreling and fewer angry words heard than might be expected in a country village. Where revolvers and bowie knives are so numerous, it is a very unhealthy atmosphere for quarreling. Thus far a strange peacefulness has pervaded the camp. Although there are probably 3,000 people in the mines, there has been no affrays, no serious fights.

> The peace and quiet which has reigned in this district since its organization is well illustrated by the fact that Frank Points, Justice of the Peace, has never had a case, either civil or criminal. He held a coroner's inquest over the remains of William Allen, but with this exception has had no official duties to perform.

1884 was a presidential year, and true to American traditions, there was political activity at Eagle right from the start.

> A meeting is called for the organization of a Republican club, Saturday, April 10th. Every Republican should remember the date and endeavor to be present. It will be held at the Republican Headquarters, or at the postoffice.

A question of jurisdiction arose in 1884, the Coeur d'Alene district being claimed by both Kootenai and Shoshone counties. The dispute was settled in favor of Shoshone in June, 1884.

Disputed Jurisdiction

> Sheriff Wm. Martin of Kootenai county is in the city, collecting licenses. Sheriff Dunwell of Shoshone county has collected the greater number, but several have paid Mr. Martin. So long as the boundary line is in dispute, parties will pay either sheriff, according to their preferences and the circumstances of the case. Judge Buck has signified to Mr. Martin that it is his intention to hold a term of the District Court in Eagle City the first week in July. Owing to the locations of the two county seats, Kootenai county would be the most convenient one for us to belong to. The general opinion, however, seems to be that we are in Shoshone county. Kootenai intends to contest the matter, and we shall soon witness the novel spectacle of justices of the peace from both counties holding court in Eagle City.

> W. H. Johnson has been appointed Justice of the Peace in and for Kootenai county, with headquarters at Eagle City. The Judge is a thoroughly informed attorney and is peculiarly fitted for the duties of his office. Frank Points is Justice of the Peace for Shoshone.

Since Eagle City was the first permanent settlement on the North Side, it established some records, such as the first burial, the first wedding, the first religious service, etc.

The First Wedding (Nugget, *April 9, 1884.)*

The first wedding in the district occurred at ten o'clock last Thursday, at the office of Singleton and Robison. The bridegroom was H. F. Scott, formerly of Shoshone Junction, Idaho, and the bride was Mrs. Laura Bell Wright formerly of Leadville, and more recently of New York. The parties have known each other for twenty years, and have been engaged for a long time. Mrs. Wright came from New York to meet her intended husband, and walked in over the Thompson Creek trail. Few weddings have more of romance than this first one in the new mines. The ceremony was performed by Rev. Mr. Shippen. Mr. J. Enright and P. Webber signed their names as witnesses to the marriage certificate drawn up by Col. Singleton. Only a very few intimate friends of the parties were present. The bride is a lady of refinement and culture, and is highly esteemed by all who know her. Both the parties are possessed of considerable means, and it is hoped they will conclude to make their home in Eagle. After the ceremonies were concluded, Mr. and Mrs. Scott returned to their residence and during the afternoon and evening received the congratulations of the friends. Several cases of very fine champagne were cracked in pledging their happiness and prosperity of the bride and groom.

Pioneer Services (Nugget, *April 9, 1884.)*

Rev. W.C. Shippen held the first religious services last Sabbath, in a large unfinished building on Eagle street. The hearers thronged the house, sitting on logs. The building could not contain the audience, and many stood in the street. Rev. Mr. Rhodes, Presiding Elder of the Oregon District of United Brethren, was present, and assisted in the performance of the ceremonies. The singing was very good, a number of fine voices mingling in this portion of the worship. The sermon was most excellent, and elicited great praise from those present. The audience listened to every word with marked attention. The hearers were from all classes, callings, occupations and professions, and may truthfully be said to have been a fair index of the representative men of the town. Members of a number of churches were present, including the Presbyterian, Baptist, Christian, Episcopal, Universalist, Methodist and United Brethren. A collection amounting to $15.10 was taken up to be used in purchasing hymn books.

Deaths

Dec. 24, 1883, William Henry Allen, aged 50 years.
Jan. 24, 1884, Malachi Johnson, aged 26 years.
Feb. 18, 1884, John Morris, aged 45 years.
Feb. 15, 1884, Charles Day, aged 35 years.

The first burial in the Coeur d'Alenes occurred on Christmas day, 1883. Wm. H. Allen had died of heart disease the day before. He had come from Virginia City, Montana, and had been a wealthy man. He was a 32nd degree mason of the Scottish rite, and he was buried with Masonic honors. He was laid in a grave on a little knoll overlooking Eagle, while snow fell silently. It was an awesome occasion.

The death of Oregon John, well known prospector, occurred recently. He had been supported by the camp in his illness. He died in sleep. ***Doc Sanborn brought out a table and placed it in front of his saloon. From this table Rev. W. C. Shippen delivered the first sermon ever preached in the new mines. It was a sermon full of touching pathos. ***It was followed by burial on the hillside. Oregon John's name was A. B. Pierce, although few knew him by that title.

CHAPTER VI

The Passing of Eagle City

Eagle City's rise as detailed in the preceding chapter was phenomenal; its decline was equally so. Within the space of a year, i.e., by the fall of 1884, Eagle's hopes had been disappointed, and instead of the metropolis of the mines she found herself only a small mining camp, with a superfluity of empty buildings.

There were three circumstances which contributed to the depopulation of Eagle City in the spring of '84. The first, and probably the chief of these, was the scarcity of mining land. The mining law of the United States (May 10, 1872) permitted placer claims of 20 acres, so located that the length did not exceed 80 rods. Although the Prichard and Eagle Creek valleys were extensive, they were soon taken up in 20 acre claims. We have seen already that A. J. Prichard himself located 80 acres in the Discovery group alone. The prospectors and miners who came early got the ground, and those who arrived on the scene later were left out. By the opening of mining activities in the spring of 1884, the area had been so completely located that not even a fractional claim could be made in the two gulches.[1] The majority of the Eagle City population were not able to locate claims. Therefore, those who did not have land and did not want to work for wages left town. Many of them went prospecting in the neighboring hills and valleys, and some of these prospectors were rewarded by rich discoveries, particularly in the lead-silver region of the South Fork. Still others left the country in disappointment and chagrin.

A second factor of importance in the decline of Eagle City lay in the mining conditions of Eagle and Prichard Creeks. When mining began in earnest in the late spring of 1884, it was found that the ground was too deep to be mined with the simple equipment at the miners' command. Bed-rock lay 30 to 40 feet below the surface of the ground in Eagle Creek and lower Prichard Creek. Furthermore, through this deep gravel ran a large flow of sub-surface water. These circumstances made it too difficult, if not quite impossible, for the average miner with pick and shovel to reach the gold deposits on the bed-rock. Indeed, it was only in recent years (1917–'26) that much of the deep ground has been worked and that by expensive and complicated dredging (see Chapter XIII), while the deepest ground, in lower Prichard Creek, has never been worked. Thus, the miners at Eagle City in 1884 found that only the side gulches and the rims of the main channels could be mined successfully.

The third cause is an extension of the second, just given. The richest and most workable mines were located three or four miles above Eagle City on Prichard Creek. This was too great a distance for the miners to travel back and forth every day. The town of Murray (then called Murrayville) had already been established in the heart of the rich diggings. Hence, there was a gradual movement of population upstream, abandoning Eagle and building up Murray.

The depopulation of Eagle City from these three causes was so complete that by the beginning of the second winter (1884–'85) the number of its inhabitants was estimated at only 150 or 200 (Aulbach's estimate) who were active placer miners or farmers. Since that time, Eagle has continued to exist first as a small mining camp, later as a center for lumbering, and finally as an agricultural region. In this last capacity, there is no town organization, the farmers living on their small tracts scattered over the two valleys.

But before its eclipse, Eagle City enjoyed a final brief period of excitement. Judge Norman Buck held a term of the territorial district court at Eagle beginning in June and continuing till August 6, 1884. At this term of court 39 civil suits were heard, the majority being litigation over mining claims, and also six criminal cases were tried. The litigation over mining claims had effectively tied up mining operations in many of the richest locations, so that the court battle was bitterly fought. Here for the first time appeared many of the brilliant members of the legal fraternity whose names have since made history in the Coeur d'Alenes.

1. *Onderdonk*, p. 123, also Chapter V. above.

The Cœur d'Alene Nugget.

VOL. I. EAGLE CITY, IDAHO, SATURDAY, MARCH 15, 1884. NO. 1.

HOW TO GET HERE.

Interesting Facts Concerning the Different Trails to the Mines.

"All roads lead to Rome," and all trails to the Cœur d'Alene. Despite the eager rivalry which exists between different stations on the road as to the nearest and most accessible point, the mines are interested in the development of all routes. Spokane Falls and Rathdrum have thus far controlled the bulk of the freighting business, goods coming by sleighs or wagons to Evolution and thence by pack trains. A very large force of men are putting a trail in from Jackass for the Spokane Falls Transportation Company. This is about completed, and will be the favorite pack trail. The North Fork and the line of steamers soon to be running from Fort Cœur d'Alene to within a few miles of Eagle, is a route from which much is expected. If the large "jam" or drift can be cut out, steamers can come to Hummell's landing, five miles from town. The "big jam" covers about ten acres, and completely covers the channel of the river with trees and rubbish. A contract was let on the 10th of January for cutting out this drift, but the work has not yet been performed. Freight can be carried up to this point, transferred to small boats above the drift, and so can come to Eagle City in high water. When the trails break up in the spring and become impassable, the river route will be the main dependence of our citizens.

THOMPSON'S FALLS

has been the favorite wagon road during the last few weeks, and men have been working to get a sleigh road finished to the mines. They are nearly to the Summit. The route possesses every natural advantage, the grades being easy and the distance shorter than by any other route. If a sufficient sum of money could be expended upon this road it would probably outrival all competitors.

BELKNAP

has the money, and may come in with the first wagon road. Over fifty men are at work a short distance from the Summit, and they are pushing ahead at a lively rate. The Northern Pacific is booming Belknap, and all that can be done will be to hasten the road to completion and make it the most popular.

TROUT CREEK

The Trout Creek route is being traveled by hundreds of people daily, and the popular trail for footmen and for carrying freight on toboggans. It is steep and rugged and difficult, and although improvements are being made by which Bald Mountain and a number of the worst places are avoided, it is a hard trail at best. Men, women and children have come over it, however; three fourths of the people in camp have arrived that way. [illegible]

THE HERON ROUTE

Thirteen of the business men of Heron, on the Northern Pacific railroad, came over the trail to Eagle [illegible]

Important Facts Concerning the Celebrated Cœur d'Alene Mines, of North Idaho.

NORTHERN PACIFIC RAILROAD CO.,
GENERAL PASSENGER DEPARTMENT.
Circular No. 6.

ST. PAUL, Minn., February 1 1884.

The wonderful rich mineral wealth of the Cœur d'Alene mountains and the tributaries of the river of that name, has been heralded to the world, and old prospectors and miners who are conversant with the history of the banner districts of Nevada, Colorado, Utah, Arizona and New Mexico stand amazed at these new fields, which are

UNEQUALED IN RICHNESS AND EXTENT, the yield being practically inexhaustible, rendering impossible any over crowding of the district by reason of too great an influx of prospectors and miners. *There is more than enough for all who come.* The minerals consist of silver, tellurium and free milling gold.

$100 PER MAN PER DAY

are being taken out of the rim rock of the gulches, while in gulches $25 to $40 per man per day are being panned out. The claims are all very rich, and are located in the gulches of the North Fork of the Cœur d'Alene river, Eagle, Pritchard and Beaver creeks, streams running into the Cœur d'Alene river. The Pritchard is the chief stream, flows two thousand inches of water, and it is said, does not freeze up, the altitude being only 2,400 feet.

$50 NUGGETS OF GOLD.

Rich placer deposits have already been discovered for a considerable distance on Pritchard creek, and the same distance on Eagle creek, the creeks being known by the latter name from the point where they come together. [illegible] On Pritchard creek can be seen valuable placer claims [illegible] correspondingly rich return in bright, golden nuggets, some of which have been sold for $50, $100, $160 and $200.

FREE GOLD QUARTZ.

[illegible]

The Timely Caution.

Strangers are constantly writing to enquire the best time to come to the mines. No human being can correctly answer the conundrum. If one hopes to become established in any business first, the time has already passed. We have fine stores, wholesale and retail liquor houses, restaurants, lodging houses, barber shops, laundries, drug stores, shoe-maker shops, lunch stands, hospitals, a saw-mill, shingle mill and a score of other enterprises either established or on the way. To be first in anything, is to make money, but it would be difficult to start any enterprise on the ground without finding a competitor. The business of buying and selling town lots is pretty well monopolized by energetic enterprising men who were here at the start and who possess every possible advantage over a new comer. The rush began about the middle of January, as soon as the Trout Creek trail was completed, and all who came brought a little money, and were eager for investments. Plenty of money is being made every day however, in town lot speculations, and anyone understanding the business can double any reasonable amount of capital within the next sixty days. Mining claims are bought and sold every hour of the day. It is positively too late to come with a hope of discovering a claim on Eagle or Pritchard creeks or their tributaries, although prospectors are finding much over ground and prospectors enough scattered through the hills at present to locate every foot of unoccupied ground. The claims which are located twice and three times [illegible] all. To purchase a claim, unprospected, one should come at once. [illegible]

"THE DAISY."

The Largest and Finest Saloon in the Mines.

COY & HESS,
PROPRIETORS.

Open Day and Night.

—Our Stock of—

WINES,
LIQUORS
AND CIGARS

Is selected with especial reference to the wants of our Customers, and everything is of

THE BEST QUALITY.

A Band of Music is in Constant Attendance.

Don't Forget to Visit The Daisy.

Nevada Chop House,

AND HOTEL.

ON EUROPEAN PLAN.

PROFESSIONAL CARDS.

H. O. BEESON, M. D. — P. B. WILLIAMS, M. D.
BEESON & WILLIAMS,

Physicians and Surgeons.

Managers Cœur d' Alene Hospital Association. Late Surgeons N. P. R. R.

OFFICE—At McNab's Drug Store.
EAGLE CITY, IDAHO.

JAMES F. TOPLIFF,

NOTARY PUBLIC, CONVEYANCER,
AND REAL ESTATE AGENT

Office—At Porter's Cabin, Eagle City

CHAS. W. O'NEILL,

ATTORNEY AT LAW.

(Corner Pritchard Avenue and Eagle St.)
Eagle City, I. T.

Mining Land Patents a Specialty.

BRADSHAW & GRIFFITTS.

ATTORNEYS AT LAW.

Spokane Falls, W. T.

L. J. FRANK.

ASSAYER [illegible]

Spokane Falls, W. T.

FRANK POINTS

MINING RECORDER

[illegible]

FRENCH, WOODS & MARSHALL

Attorneys at Law.

Offices: Eagle City and Murray

W. M. SCALLON,

ATTORNEY AT LAW

Eagle City.

[illegible]

ATTORNEY AT LAW

[illegible]

PHYSICIAN AND SURGEON

The Coeur d'Alene Nugget, Volume I, number 1.
(photostatic copy)

Delta, Idaho, in 1890.

CHAPTER VII

Murray

The new and permanent center of the North Side of the Coeur d'Alenes was Murray, located on the north bank of Prichard Creek between Alder and Gold Run gulches. The town was founded January 22, 1884, as Murrayville, or Murraysville, and took its name from the fact that it was built on the Pacific placer claim owned in part by George Murray. An account of Murray's progress in its very early days appears in the *Coeur d'Alene Nugget*, March 15, 1884.

Murrayville

This new and enterprising town is situated in the central portion of the mining district, about half way between Raven and Eagle. It is at the famous Widow claim and the best developed mines are in its immediate vicinity. The saw mill is to be erected two miles away. And some of the finest quartz ledges yet discovered are within a short distance of the town. Murrayville is laid out into regular streets on the banks of Prichard Creek on what is known as the George Murray claim. The site is one of the most beautiful and romantic imaginable. Timber is plentiful and building is going on as rapidly as in any part of the diggings, which means that every man is at work. J. H. Miller is City Recorder, and Judge Wm. H. Claggett is Recorder of the Murrayville Mining District. The boys donated the Judge a fine lot, and built him one of the best houses on Prichard Creek. They have given Dr. Lansing a lot on which to erect a hospital, and seem to have a ready spirited manner. Their faith in the permanence of their city is based upon the fact that it is in the very heart of the mines close to the quartz ledges, and necessarily near the place where hundreds and thousands of miners will find employment. A brisk business is being driven in town lots, prices varying from $100.00 to $1000.00.

Murray's growth was rapid. It received a great portion of the emigration from Eagle City. In addition, early settlers and later arrivals combined to swell the total number of inhabitants to not less than 2500 by the end of the summer of 1884.[1] Murray had a cosmopolitan population, almost every nation of the civilized world being represented (with the exception of Asiatic races; see Chapter IX); and all the famous old mining camps of the west had their representatives there too. In 1885, while the boom was still on, Onderdonk describes Murray briefly as follows:

> Murray is now the business center of this mining region; many substantial business houses have been erected here, fire-proof cellars provided, in which tons of goods are stored for the trade of the country, and a number of very comfortable dwellings have been built, and a commodious hospital is being erected for the care of the sick, a public school has been established, a good bank has been opened, and a live newspaper, the *Idaho Sun*, is published tri-weekly. Every branch of business is carried on in the spirit of genuine business enterprise. The medical and legal professions are well represented, and Murray seems favored with all the elements that contribute to the permanent prosperity of a mining town. The people of this great mining district are of no ordinary class, but are of more than average intelligence, who see first the great resources of the West, and with comprehension and earnest spirit of enterprise seize and appropriate them as the rightful inheritance of the industrious and self-sacrificing citizens of this free domain for themselves, their heirs, and their country.

An excellent photograph which antedates 1890 shows Murray in its early splendor, and a map of 1894 shows the city in outline.

The building of wagon roads to Murray in 1885 ended the isolation of the North Side communities. In the spring and summer of that year, roads were completed between Murray and Thompson Falls, Montana, and between Murray and the Old Mission, the latter by way of Osburn and Two-mile Creek. The first stage into Murray from the West arrived from Spokane Falls July 28, and the first one from the East came in from Thompson Falls, November 16. Another wagon road over Dobson Pass and Nine-mile canyon connected Murray with Wallace in 1890. Telephone connection between Murray and Thompson Falls, established in May, 1886, made communication still easier. From that time on, daily mail service, regular stage and freight lines, and the news of the world made their way comparatively swiftly into the valleys of the Coeur d'Alene district.

County Seat Agitation

Shoshone County was organized July 8, 1861, as a division of Washington Territory, with Pierce City as county seat. With the organization of Idaho Territory in 1863, the county government was transferred to the new territory without change of seat or of personnel. Just previous to the great rush into the Coeur d'Alenes, i.e., in 1881, the white population of the southern part of the county, the region around Pierce City, had dwindled to less than a hundred, and the county organization was falling to pieces. There was a determined move afoot to annex Shoshone County, or at least that inhabited southern portion, to Nez Perces County.

With the rush of 1883 and '84, however, the situation altered immediately. The population jumped into the thousands, and a clamor arose for a county seat closer than the remote and almost inaccessible Pierce City. Eagle City made a bid for the county seat, but before any action could be taken, Murray became the center of activity. The plea of the Coeur d'Alene residents as well as the Nez Perces annexation proposal both went to the Idaho territorial legislature in 1884. The resulting act provided that the seat of Shoshone county be moved temporarily to Murray, and that a special election be held in Shoshone County the first Monday in June, 1885, to determine the permanent seat. The bill was signed by the governor in

December, 1884. In the election June, 1885, Murray received 1075 votes out of a total of 1537 votes cast, and became officially the county seat. The county government had already been set up in Murray in the spring.

Thus, from 1885 until 1899, Murray was the political capital of the Coeur d'Alenes. The political contests, the sessions of the district court, the activities of a great number of lawyers, many of whom were men of outstanding ability, all lent color to the local scene for many years. By an election in 1898, the county seat was removed to Wallace (actual removal in 1899), for by that time the bulk of the population and the great mining interests were centered on the South Side of the Coeur d'Alenes. Still another political change occurred in 1904 when the southern part of Shoshone County finally received its independence and was organized as Clearwater County, with Orofino as its seat.

Adam Aulbach and the *Sun*

No discussion of Murray could be complete without mention of Adam Aulbach, pioneer journalist and enterpriser. Following a vivid career in Montana, California and Nevada, Aulbach came to the Coeur d'Alenes in the spring of 1884, with a complete printing outfit. Finding it impossible to carry his outfit over the Bitter Root divide until the snow had receded, he located temporarily at Belknap, Montana, and began issuing the *Belknap Sun* as a weekly paper, carrying red hot news from the mines across the divide. Aulbach appeared in Murray on April 14, 1884, and after looking over the prospects decided to bring his *Sun* plant into the camp at the earliest possible opportunity. This he accomplished by the aid of a train of pack mules. The first issue of the *Coeur d'Alene Sun* appeared at Murray July 4, 1884.

From that date on, for nearly 25 years, Aulbach edited and published the *Sun* at Murray, first as a weekly, then a tri-weekly, later for a time as a daily, and finally again as a weekly paper. The *Sun* was strictly a local paper, very complete in its scope, and it was a fine example of personal journalism. The few copies in existence give a clear cut picture of local happenings, and many earlier historians quote it as a source. The complete files of the *Sun* were destroyed in a fire in 1918, and this priceless fund of information was thus lost forever.

Mr. Aulbach, however, has remained at Murray, and his own private records and his own keen mind still furnish the researcher with valuable information and amusing sidelights on the pioneer days in the Coeur d'Alenes.

Economic Life

The basis of Murray's economic life, of course, was gold mining. Placer miners and quartz mill workers made up a large portion of the population. All other enterprises depended directly or indirectly upon the mines. Stores of all kinds, banks, restaurants, hotels, stage lines, freight companies, breweries, saloons, theatres, and other commercial or recreational ventures throve in Murray through the 1880's and '90's, all prospering or failing generally in direct ratio with the fluctuations of the mining activity.

Social Life

Murray boasted a rather lively social life in the days of her supremacy on the North Side. The population was of a cosmopolitan nature. Indeed, it had almost a foreign aspect, for the French and Italian tongues were spoken by many.[2] By the fall of 1884, many miners' wives had come into the camp.

Comfortable log houses had been built, and dances with good music were frequent occurrences.

Religious activities began early at Murray. The first Catholic mass was celebrated Sunday morning, November 2, 1884, by the Rev. Louis Jacquet, a Jesuit missionary. The first Episcopal service was held Sunday, October 25, 1885, in the old court room, with Bishop Daniel Sylvester Tuttle presiding. The Episcopalians of Murray later built an attractive church building which now stands in a state of extreme dilapidation. The Methodists also built a church (1896), but they located it against a hillside where a slide wrecked it. The Methodist minister, George W. Coplen, was the owner of the Daddy lode, a quartz mine.

The fraternal societies flourished at Murray. The Ancient Order of United Workmen lodge was organized first, on November 24, 1885. The Masons organized the Coeur d'Alene Lodge Number 20 in November, 1886. The Grand Army of the Republic mustered a post in October, 1886, which at one time numbered 72 members. The Odd Fellows and the Knights of Pythias were also represented among the pioneer organizations in the Coeur d'Alenes.[3]

Adam Aulbach speaks of a school in Murray in the fall of 1884, with Lucy Dean as the first teacher. The *History of North Idaho* describes a private school established in September, 1884, under Mrs. Robert Neill, and a public school begun in March, 1885, with 25 pupils under Miss Selma Talbott. In any case, organized education began in Murray early in the town's history. In 1897, a large wooden school building of four rooms was erected. It is still in service, although only one of the rooms is now in use.

Of intellectual activities other than schools, we have already mentioned the Murray *Sun*, edited by Adam Aulbach, a leader of public opinion. Aulbach had a bitter personal rival in the *Evening Record*, edited by a certain Culver.[4] Then there were the lawyers. Aulbach declares that the membership of the Murray bar was probably the most brilliant ever assembled in a mining camp. William H. Claggett, known as the "Gray Eagle" of the mining frontier, was outstanding among the lawyers from the very beginning of the community. He had been prominent in California, Nevada, Montana and Dakota before coming to the Coeur d'Alenes. Later, in 1889, Claggett rendered a great service to Idaho when he presided over the constitutional convention which preceded the state's admission to the union. Major W. W. Woods was another great character; he had been a Civil War officer, and he was to be a beloved judge of the local district court for many years. W. B. Heyburn, later United States Senator from Idaho, Alec Mayhew, Frank Ganahl, W. Y. Pemberton, Albert Allen, Tom Singleton, "Aztec" Stanford, Bob McFarland, Charles O'Neil, and William T. Stoll were also among the pioneer lawyers of Murray.[5]

In contrast to the progressive and enlightened groups just described, Murray possessed an underworld of intense vitality. The population of Murray in 1885 was estimated at 1000 to 1500 (Aulbach's estimate), yet the town supported 45 licensed saloons! There were also four dance halls in operation that year. Gambling of every character existed in the camp, generally in connection with the saloons. A red light district operated in Gold Street, in which Molly B' Dam' and May Lord and other famous "queens" reigned. In these respects Murray showed itself a typical western mining camp.

Christmas, 1884

An interesting conclusion to this description of Murray's social life is the account of the first Christmas in the camp.[6] Many families had settled in Murray, and there were about 75 children in the camp and its vicinity, so that it would be a real Christmas compared with the exclusively adult male holiday of the previous year at Eagle City. A cold wave hit the valley, depressing the thermometer to 23 degrees below zero just before Christmas. Placer mining came to a complete stop. The temperature moderated a little and a heavy snowstorm set in which lasted all of Christmas day.

There was good cheer in the camp. with universal good feeling. The miners congregated in town and made merry, the saloons doing a holiday "rush" business. Children romped and played in the snow before the log houses from the chimneys of which poured columns of smoke indicating warmth and cheer within. Houses were decorated with evergreen branches sprinkled with fresh snow.

At the Christmas dinners the traditional fowls were absent, but there were plenty of other delicacies at hand, at reasonable prices too, such as deer and bear meat, and even oysters—the last shipped in by sleigh from Thompson Falls.

William T. Stoll tells how the term of district court adjourned on Christmas Eve, after the snowstorm had set in, which made travel impossible, and left Judge Norman Buck and all connected with the court snowed in. So, making a virtue of necessity, a grand Christmas banquet in honor of Judge Buck was given at the Louisville Hotel on Christmas night. Lawyers, court attaches, mining men, business men, all of the leading citizens of Murray were there. The banquet began at 8 o'clock in the evening and lasted till 3 o'clock the following morning.[7]

Law and Order

In the chapter on Eagle City, I quoted several contemporary statements to the effect that Eagle was a peaceful camp, comparatively speaking. The records indicate that Murray was likewise a reasonably quiet and law-abiding community. Indeed, the claim has frequently been made that there was less major crime in Murray than in any large mining camp. Homicides were not unknown, but they were largely confined to the underworld. Organized lawlessness such as had existed in the Montana and other Idaho camps never vexed the Coeur d'Alene communities, or at least only once.

That single instance is related by Adam Aulbach.[8] He tells how, late in 1884 and in '85, the O.K. claim which lay just below Murray became the headquarters of a gang of young rowdies who spent their idle moments practicing with rifle and revolver, sometimes rather regardless of bystanders. Their visits to Murray were occasions for brawls and for shooting-up the town. Their leader was a young placer miner named George Bell, 25 years of age, who was an accurate marksman and very quick on the trigger. The story goes on to tell how George Bell was eliminated from the leadership of the gang after being shot in the shoulder while exhibiting his prowess at twirling an automatic pistol on his right forefinger, firing at each revolution. Bell was disabled and was sent East for treatment. He never returned. "Kid" Price succeeded Bell as leader of the gang whose activities grew more intolerable to the community. The result of this development was the formation of a "Citizens Committee," of over 50 men, whose membership was at the time kept secret, and has always remained so. The committee served notice on the "Kid" who soon afterward vanished, leaving the gang disorganized. The final item here is the amusing suggestion that when the Yukon Gold Company's dredge worked over the O.K. claim, it made a great cleanup of lead.

Decline

Murray's golden days are past and gone. She stands today a gaunt skeleton of her former grandeur, harboring about 100 people and many empty buildings in her constricted area. The story of her decline is a melancholy one.

The boom of 1884 continued strong through '85. But by 1886 the placers began to weaken. Quartz mining sustained the town for a number of years, but its course also proved to be brief. By 1900, all the quartz mills had shut down, and the decline was on it earnest.

During all these years, even in the years of the greatest gold mining activity, 1884 and '85, the discovery and development of the rich lead-silver deposits on the South Side of the Coeur d'Alenes had been going on steadily and rapidly. Wardner, Burke, Mullan, Wallace, Gem, Kellogg, and others—these were the big camps of the Coeur d'Alenes in the 1890's and later. The fascinating history of this mining development lies outside the scope of this study. Suffice it to say that the population of the North Side was gradually drawn off to the South Side.

Murray saw a brief revival of prosperity early in the present century. The Idaho Northern Railroad entered the Prichard Creek country in 1908.[9] Also, several big lead-silver mines were discovered and began to produce on the North Side, such as the Jack Waite, the Monarch, the Beartop and the Paragon. These enterprises brought a new lease on life to Murray. But in January, 1918, a flood washed out the railroad in Prichard Creek above Eagle, and it has never been rebuilt. The North Side lead mines, including even the Jack Waite, which is by far the most promising, have all closed down, some of them permanently.

Finally, in 1922, the Yukon Gold Company's big dredge, after working Prichard Creek downstream from Murray, returned to the townsite and proceeded to eat up a large portion of it. All the Prichard Creek frontage and all of Gold Street were dredged, the buildings being torn down and unsightly piles of boulders left on the premises. The boulder piles extend clear up to Main Street in a few places, and they fill the back yards of all the houses on the south side of Main Street.

Thus, Murray lives on in the twilight of an interesting and romantic past. Gold brought about her foundation; much gold still remains in the ground around her, and some day it may cause a revival of activity in her now quiet streets.[10]

1. Aulbach's estimate.
2. Conversation with Adam Aulbach, August 1, 1931.
3. *History of North Idaho*, p. 1043.
4. Stoll's *Silver Strike* is the authority for the *Evening Record*. No copies are known to have survived.
5. William T. Stoll gives an interesting characterization of these men in *Silver Strike*, pp. 17–19.
6. Aulbach: *The First Christmas in the Coeur d'Alenes*.

7. *Silver Strike*, pp. 38–39.
8. Aulbach: *Reminiscences of Prichard Creek, Idaho.*
9. The Idaho Northern Railroad, a subsidiary line of the Union Pacific System, built its line from Enaville (on the Spokane-Wallace branch) up the North Fork of the Coeur d'Alene River to Prichard and thence up Prichard Creek to Paragon Gulch. The line was built to reach valuable timber stands which were being logged off as well as to serve the mining community, and in the former capacity it has been and is still a valuable link for the North Fork area.
10. Apropos of Murray's present stagnation is the remark attributed to Mr. Vivian Green of New York, sole owner of the Golden Chest Mine: "I would rather lie dead on Broadway than live in Murray." Aulbach: *History of the Golden Chest Mine.*

Murray, Idaho, in 1890.

Murray, Idaho, in 1931.

Looking West on Main Street,
Murray, Idaho.

Adam Aulbach, of Murray, Idaho.

Old County Court House and County Jail,
Murray, Idaho.

Fraternity Hall, Murray, Idaho.

Abandoned Railroad Station,
Murray, Idaho.

Episcopal Church, Murray, Idaho.

High Bench Gravel Excavation, in the Old Channel Wash, near Murray, Idaho.

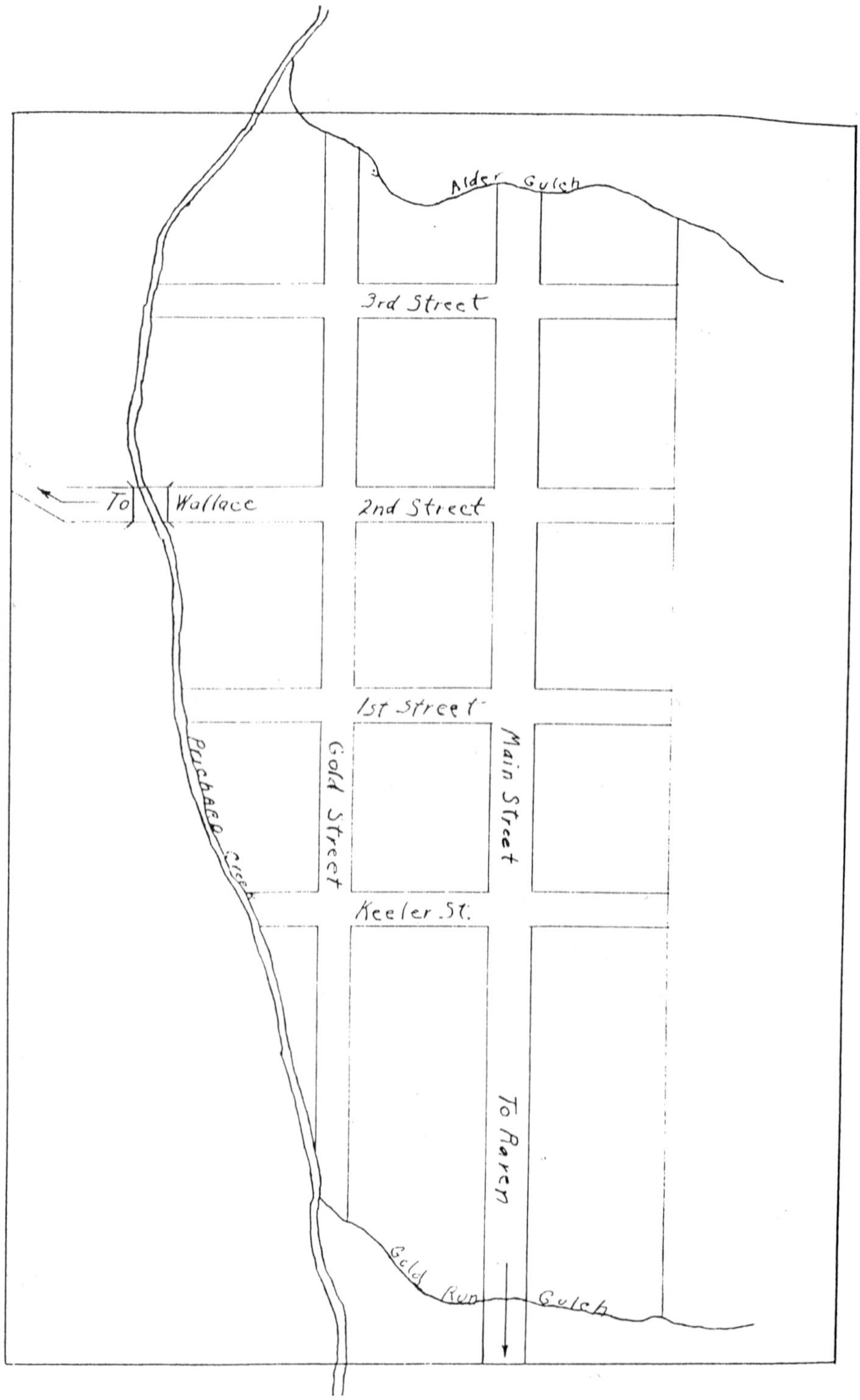

Map of Murray, Idaho.

CHAPTER VIII

Other Camps

Eagle City and Murray, although they tended to monopolize the story, were by no means the only camps in the gold fields of the Coeur d'Alenes. On the contrary, in the boom years, 1883 to '85, every little mountain prairie along the main routes of travel supported a collection of tents and houses containing a small but ambitious populace loud in proclaiming that location as the real center of the mines, and the future metropolis of the whole region.

Delta

Of these minor towns the most imposing was Delta. Fortunately located on Beaver Creek, at the mouth of Trail Creek, at the convergence of the two main trails to Murray and Eagle, Delta became a settlement of importance before the close of 1883. Its population at its peak has been estimated at 1000 to 1500.[1] It had the reputation of being a real western mining camp, with almost one fourth of its buildings occupied by saloons, dance halls and gambling houses. In the county seat election of 1885, Delta was Murray's chief competitor for the honor, and polled 459 votes. As long as the placers of Trail and Beaver Creeks and their tributary gulches continued to operate profitably, Delta prospered. But in the 1890's the decline set in, until Delta became known chiefly as a point at which horses were rested or changed on the Murray stages. The death blow of the community was finally struck in 1902 when the townsite was acquired by the Mascot Gold

Mining Company who began operating a dredge over the ground. The buildings were torn down and in their place appeared disorderly heaps of boulders when the dredge had passed. At the present there is nothing to be seen at Delta save the cross-road sign pointing the ways and telling the distances to Murray, Prichard and Wallace.

Littlefield

Two miles above Murray on Prichard Creek, at the mouth of Butte Creek, lay Littlefield, which was first known as Butte City. In its final name it honored Dr. H. L. Littlefield, one of the pioneer claim owners of the townsite. The *Coeur d'Alene Nugget* on March 15, 1884, describes the settlement as follows:

Butte City

> At Butte City, February 25, a meeting of citizens was held, and it was resolved that in order to secure titles to town lots, foundations four feet high must be laid within ten days, the lot must be recorded by March 22d, and a cabin 10 x 12 must be built before May 1st, or $50 worth of improvements must be made. No one can hold more than two lots. Fifteen blocks were surveyed and located, and many locations have been made outside the survey. A block was located for a Court House and School house. A number of business houses are going up rapidly, and the camp is booming. Lots cost from $500 to $500, and are increasing in value almost daily.

Adam Aulbach mentions Littlefield as "a booming camp of 25 houses" in April, 1884. Littlefield's ambitions, however, were doomed to disappointment, and in a few years it passed out of existence, leaving only a couple of faded and dilapidated house fronts to mark its grave.

Raven

Raven was a promising little community in 1884, and also in '85 and part of '86. It was situated on Prichard Creek at the mouth of Bear Gulch and close to the mouth of Granite Gulch. The same early issue of the Coeur d'Alene Nugget describes Raven thus:

Raven City

> This enterprising town was founded January 19th, Messrs. Kennedy and Riarson driving the first stakes. When the Thompson's Falls road comes, Raven, is bound to become a prominent point. At a citizens' meeting held there March 1st, it was agreed that $100 worth of improvements should be placed on each lot prior to May 15th. Two days work on the trail to the Summit entitles lot owners to a credit of $10 upon their town lot improvements. Some of the richest gulches, including Bear gulch, come in close to Raven. Mr. D. Kennedy has made a survey and plot of the town, and lots are bringing fair prices.

For a while in 1885 and '86, after the Thompson Falls road had come through, Raven prospered. It had a bank, a number of business houses, with the usual high percentage of saloons, eleven of them in fact. But one day in August, 1886, a catastrophic event occurred. A group of men at Murray were clearing a field for a baseball diamond. They set afire a pile of brush and other refuse. The rest of the tale can easily be imagined. The fire got beyond control and swept up into the forest, very dry at this season. The vagaries of the wind carried the flames over the slopes on the south bank of Prichard Creek; Littlefield was spared, almost miraculously; but Raven was burned to the ground, and never rebuilt. The extent of this disastrous fire can still be traced in Tiger Gulch south of Murray, where a clear-cut line divides the old timber from the new growth.

Robert Wayne Smith

Thiard (Myrtle)

One more camp and the story is complete. The last is Thiard, called Myrtle for many years, which stood on the south side of Trail Creek between Potosi and Placer Gulches, a little over a mile above Delta. In the years 1885–'86, Myrtle counted close to 1000 inhabitants and it is said to have handled more gold dust in those years than any other camp in the region. Now, however, all that remains is a small cluster of old buildings in varying stages of wreckage. The only ones in use seem to be serving as barns for a herd of cows.

1. Aulbach: *The North Side of the Coeur d'Alenes*, 1928.

CHAPTER IX

No Chinese!

One of the distinctive features of the Coeur d'Alene district is the complete absence of Chinese labor, or indeed of any oriental residents. Chinese miners and laborers have lived in the Clearwater and St. Joe regions to the south, and in the Clark Fork Country to the north, but there have never been any Chinese in the Coeur d'Alenes, and in all probability there will be none for many years to come. This tradition, still potent today, had its beginning in the earliest years of the community.

In California, and in many other western mining camps, including even Pierce City in the southern part of the same county (Shoshone, now Clearwater), the coming of Chinese miners has spelled the ruin of the community. The orientals, content to work for smaller returns, have come into the placer mines after the first and richest diggings have been exhausted. They work the less satisfactory ground and often rework the claims which white men have finished and abandoned. The Chinese have generally done very thorough work, and after they have finished there is seldom anything to be taken from the ground.

The very first suggestion of Chinese miners in the Coeur d'Alenes aroused a storm of protest and opposition. In the *Coeur d'Alene Nugget*, Volume I, Number 2, March 22, 1884, there appeared a significant editorial which explains completely the local feeling at Eagle City.

No Chinamen Need Apply

We want Chinamen, but we want them a long way off. As cooks they would oftimes be acceptable; as laundrymen they would be convenient; as pack animals on the different trails they would be bonanzas. But they had better stay away just the same. The fiat has gone forth that no Celestial can ever gaze on the wooded gulches of the Coeur d'Alene and live. This camp, like Leadville, will never feel the curse of coolie cheap labor. There is but one voice upon the subject. A man who would advocate Chinese immigration would be snubbed. The first representative of the Flowery Kingdom will meet with a reception which in all probability will satisfy his countrymen that the mines won't pay. John Chinaman got into the California mines, into many other mines, but he must not think of attempting a visit to those of Northern Idaho. If he insists upon coming, however, let him bring a roast hog, plenty of fire-crackers and colored-paper, and all the essentials for a first-class Chinese funeral. He needn't bother to bring the corpse. It will be in readiness. Ta! Ta! John.

The same attitude in regard to the Chinese was expressed three years later at Murray. Adam Aulbach tells of the incident in the following words:[1]

"Gradually, as the placers began to weaken on the rims, an agitation developed in 1886, and became dominant in '87 when claim owners made an effort to introduce Chinese labor. This agitation, however, was very strongly opposed by the laboring element in the placers and by other occupations. This resulted in a public meeting July 4, 1887. A platform was erected in a public street in the upper part of town and practically every man living on the north side was present. The meeting was addressed by representatives of the claim owners and also by laborers and by the public in general opposed to the introduction of Chinese labor. Among the speakers were Phil O'Rourke and Adam Aulbach.

"The determination upon the conclusion of the speaking was against the introduction of Chinese in any form or conditions. A vote was taken, and the crowd almost to a man moved to the north side of the street at the request of Mr. Aulbach, thus demonstrating that the opposition to the Chinese was practically unanimous. The south side of the street was represented by claim owners and attorneys.

"This was the final test of the feeling of the North Side of the Coeur d'Alenes, which naturally also was adopted by the miners of the South Side. No Chinese have ever been permitted to work in the Coeur d'Alenes.

"A contingent of Chinese laborers were brought into lower Prichard Creek the year following to work the placers there, but they were immediately served notice by a committee of miners from upper Prichard Creek to leave the country. This notice was final and the Chinese left. Since then none have entered and there has been no further question."

A story persists in the Coeur d'Alenes telling of the hanging of two Chinamen who came into the region in defiance of repeated warnings such as those quoted above. Two-mile Creek, north of Osburn, the old road to Murray, is the locality generally ascribed to the lynching. All research, however, has failed to reveal a single written reference to the incident, and the direct questioning of a number of "old-timers" in the Coeur d'Alenes likewise has failed to uncover any facts. So persistent a story, however, must have some foundation. Therefore, I offer the following incident as a possible explanation:[2]

"The one great tragedy of 1885 in Shoshone county occurred at Pierce City in September. On the 10th of that month at an early hour the body of the well known and respected old pioneer, David M. Fraser, was found in the deceased's own store, literally hacked to pieces with axes, hatchets and knives. A bullet, also, had entered the mouth and passed out through the side of the neck. An old county safe had been battered and the lock broken, but Mr. Fraser's own safe was untouched, as was all the rest of his property. The body was taken to Lewiston for interment. After its arrival there twenty or twenty-five citizens left for the scene to endeavor to find a clue to the murderers and these were joined by about fifteen from Camas prairie. Upon their arrival they surrounded the town, sending in a small posse to learn of the situation. It was found that the few white men in camp had arrested eight Chinamen on a charge of complicity in the crime, including two Chinese merchants of Pierce. A preliminary hearing had been given them and five were held for trial, the others being discharged. It developed that the crime had been committed for the purpose of ridding the town of the only white merchant, that the Chinamen might have a monopoly of the business, also that the Mongolians were angry at Fraser for taking the part of Indians against them when they had paid the red men in bogus gold dust.

It is said that the evidence upon which the citizens held the Chinamen was obtained by disguising as an Indian a young man named Sears, who could speak Chinese. Sears stole up to the courthouse jail and heard the eight Chinamen incarcerated there discussing the crime and what were best to do to escape punishment.

After the departure of the Lewiston and Camas prairie people the few white men left began asking one another what was to be done with the accused Chinamen. It being evidently impossible to guard them in Pierce until next term of the district court, the whites decided to take them to Murray. Seven men, one of whom was the deputy sheriff, started with the prisoners. When they were about four miles out they were met by a mob, only one member of which was masked. The mob overpowered the guard, seized the Chinamen and, notwithstanding the fervent pleadings of the wretched Mongolians, hanged them to a pole between two trees. No punishment was ever meted out to the lynchers, although the matter was officially investigated by Secretary of State Bayard in July, 1886, for the purpose of satisfying the Chinese government.

Another incident, which may or may not be true, at least typifies the present day antipathy to Orientals in the Coeur d'Alenes. They tell a vague story, unsubstantiated in print, that the private car of a Northern Pacific official lay in the station yard at Mullan. A Japanese cook, employed on the car, ventured into the business section of the town to buy some groceries. He was met with organized resistance, and the fire department chased him back to the car with the aid of a stream from the fire hose. Whether true or not, as I said before, this tale represents the feeling of the Coeur d'Alene communities on the question of Orientals, and I have noticed that Oriental tourists, including Filipinos, drive hastily through the region without making any unnecessary stops.

1. Conversation, August 1, 1931.
2. *History of North Idaho,* pp. 994–995.

CHAPTER X

The Placer Diggings

The placer mines of the Coeur d'Alenes were really rich. The Northern Pacific's gold circular did not do great violence to the facts in that regard. The weakness of that interesting document lay in the overstatement of the extent of the gold fields. They were limited in area.

Twelve miles on Prichard Creek, from Raven to the mouth of the creek, with 17 tributary gulches; the Old Channel Wash, for 6 to 8 miles, paralleling Prichard Creek on a higher level—a phenomenon which I shall describe later; Eagle Creek for 10 miles on both main forks, and the tributary gulches; Trail Creek for two miles with its tributaries; Beaver Creek from Pony Gulch to the North Fork, about 4 miles, with its tributaries. These were the placer areas of the Coeur d'Alenes, and a brief study of the map of the district will show how limited the total area really was. Then, subtract a large portion of the ground which was too deep to be worked by ordinary mining methods, and another portion where the water supply was inadequate, and you have an active placer area very restricted in size.

The mining began in the spring of 1884. On some favorably located claims they were digging early in April, while on others no work could be done until June, after the spring freshets had receded. Still others were tied up in litigation until August. But, generally speaking, the spring and summer of 1884 were feverish with mining activity.

Prichard Creek Area

On a unknown date, probably early in the spring, a meeting of miners took place at Eagle City. Here the Coeur d'Alene Mining District was organized and a set of by-laws, thirteen sections in all, was drawn up. These by-laws, which were published in the *Coeur d'Alene Nugget* April 9, 1884, are repeated in full in Appendix VI. They set up definite limitations on the size of claims, and definite regulations for assessment work, recording of claims, use of water, and other necessary arrangements. It was the customary procedure for each mining district to follow, yet this set of mining laws is the only set drawn up in the Coeur d'Alenes which has been preserved.

As we have mentioned before, the main channel of Prichard Creek was too deep to be mined easily, so that the work of 1884 was confined to the rims of the channel and to the tributary gulches. Of the latter there are 17, all entering Prichard Creek from the north side between Eagle Creek and Bear Gulch. The larger gulches are named on the map, e.g., Buckskin, Alder, Dream, Accident, etc., but many, including some of the richest, such as Missoula, Dry, Nugget, etc., are too small to be drawn on a map of this scale. Of the 17 tributary gulches, Dream, Buckskin and Missoula were the richest. The rims of Prichard Creek, were cleaned up and the tributary gulches worked intensively in those early years. Hundreds of thousands of dollars worth of gold were taken out, chiefly in nuggets and coarse grains. The deep ground in the main gulch remained untouched.

Let us trace the progress of the placer mining. The *Coeur d'Alene Nugget*, April 9, 1884, at the beginning of the season comments as follows:

The Placer Diggings

It is impossible at the present time to give much authentic information relative to richness of the placer diggings, The "Widow" has unquestionably yielded from $12 to $20 per day to the man. The aggregate amount of dust taken from the claim exceeds $18,000. The Peter's claim only ran about three weeks, and cleaned up from ten to twelve thousand dollars, according to the statements of Dr. Campbell who was in a position to know the facts, W. E. Cole saw one lot of dust weighing $800, which the parties asserted had been washed out by three men in three days, the work having been performed in six inches of water. Eleven claims have lines of sluices and are said to be down to bed rock. These are the Widow, Ives, Happenstall, Wyant, Macomber, Campbell, Points & Co., McQueen, Moscow, Geo. Murray, Rockford Co., and Allman & Co. On all these claims, and many others, work is now progressing. Drainage ditches are being cut, and sluices are being placed in position. The "Widow" claim is now in shape for mining, and pay dirt is being washed. On Bear Gulch and all the numerous tributaries of Prichard Creek, the miners are busy at work. A few weeks will solve the problem fully and gold dust will either be plentiful in the various camps or else the mines will be voted a failure so far as placer diggings are concerned. No fears are entertained regarding the result by those whose money is invested. On the older claims it is boldly asserted that the question has been fully decided, and the answer is "there's millions in it."

Frequent mention is made of the Widow claim. This rich ground was unquestionably one of the banner producers of the region. In the same issue of the *Nugget* appears this description of the mine:

The Widow

Last Week's Clean-ups aggregate $1,160.00.

Everywhere that the Coeur d'Alene mines are known, the fame of "The Widow" claim has extended. All kinds of strange romantic stories are circulated regarding its history, its wealth, and the persons interested in its ownership. "The Orphans" is the name usually given to the nine men who have possession of the mine. They are very quiet, gentlemanly fellows, who mind their own business, have little to say, and yet are always courteous and polite. While they do not care to make their business public, they never object to anyone watching their clean-up at the close of the day. Anxious to get every possible fact regarding the true value of the district, we visited the mine last Wednesday. Thanks to Mr. Keeler, the superintendent, and to the gentlemen associated with him, every attention was shown us, and every opportunity given us to see and learn the value of the mine. We picked gold from the crevices of the bed-rock, and were taught where to look for pockets and rich spots. The plan of working the mine, the difficulties against which they are now contending, and all information requested was freely given and explained. We shoveled dirt into the sluices, picked up pieces of gold and dropped them into the boxes to see if they could be found in the clean-up, and enjoyed an exceedingly pleasant afternoon. There were only three men throwing dirt into the sluices Wednesday, the other five being engaged in stripping ground for future work, or attending to other duties. The clean-up for the day was $130. The gold was very coarse, but there were few of the large nuggets which were so common in the ground worked last fall a few yards up the creek. The latter part of this week the boys expect to strike this rich ground again, when the larger part of the gold will consist of nuggets worth from five to twenty-five dollars. The clean-up at night, last fall, used to run from $200 to $800. Over eighteen thousand dollars was taken out by the boys, and it all came from a very small strip of ground which is supposed to be in the channel of an ancient river. At present, the work performed is with a view of preparing a dump for the richer ground above. Many disadvantages are encountered now which will be overcome in a few days, yet the clean-ups of the past week would be considered quite satisfactory in most claims. Monday was representing day up the gulch, and little work was done on the Widow that day, and no clean-up was made. On Tuesday $158 was cleaned up. The yield of the Saturday previous was $160. Thursday and Friday's run was $450, and Wednesday's $270. Nine men are interested in the claim, but only eight are in the mines at present. The aggregate amount of gold taken out from Wednesday to Wednesday inclusive, was $1160. Only six days work was really performed, which gives over $24 per day to the man.

The history of this extraordinary claim is as follows: The Widow was originally located for Mrs. M. H. Lane by A. J. Prichard, her attorney in fact, March 24, 1883, and was relocated and recorded June 11, 1883, on page 36, Book A. records of Coeur d'Alene Mining district. In September of last year Mr. Prichard had a number of men working under contract to develop the claim for Mrs. Lane. Each night when a clean-up was made, it is said that one fifth was scrupulously weighed out and set apart as the "Widow's Mite" and given to Mr. Prichard as her agent. Mrs. Lane is the widow of a man who was a comrade of Prichard during the war. She lives in Illinois. Mrs. Abe Edgington, who has been spending the past few weeks at Fort Coeur d'Alene, has generally been supposed to be the widow of the Widow's claim. This, Mr. Prichard says, is only an idle rumor, Mrs.

Edgington being an entirely different personage. The gentlemen who now hold possession of the claim, were prospecting in the gulch last summer, and became satisfied that this claim could be jumped. One of their number located the ground, and the party came and asserted their rights by turning off the water. No resistance was made by those working the claim, but a day and a half was asked and granted in which to clean up the sluices. The claim was then abandoned to the jumpers. There was no force employed, no threats, not even a harsh word. The Orphans believe they have just legal right to the ground on account of irregularities in the title of Mrs. Lane. They were strangers to each other until they met in the Coeur d'Alenes. The claim is situated in the upper portion of Murrayville, and is one of the greatest attractions in the district. It is close to the trail, and almost every passer-by stops and watches with eager interest the miners at their work. From one of the gentlemen who worked the claim last summer we learn that about $5,000 was taken out during the month preceding the jumping.

Incidentaly, the question of title to the Widow claim was settled in favor of the jumpers at the summer session of the district court held at Eagle City, June to August, 1884.

A month later interesting reports continued to be given out by the *Nugget*. From the issue of May 7, 1884:

"All the mines have yielded finely this week. The George B. Ives, H. E. Wolf and the Widow continue to yield over an ounce a day to the man. Missoula gulch lacks water, but is recognized as ranking among the best paying claims. Lucky gulch and the Gelatt are getting ready to turn out handsome quantities of dust.

Gen. A. P. Curry has interviewed the bankers and business men of Eagle and found that exactly 58 ⅓ pounds of gold dust were handled by them in the last two weeks. (over $18,000) This is doubtless only a small part of the dust taken out, most miners preferring to keep the dust themselves.

Buckskin gulch has given the camp the principal excitement of the past week. Hundreds of men have flocked thither to witness the unusual spectacle of men picking gold from the earth instead of washing it out in a pan. This is about what they witnessed. Robert Williams, David Bryant, John F. Hennessy and Frank Barnes own claim no. 3 and upon striking bed-rock got from $2 to $20 to the pan. The entire bed-rock averaged $5 to the pan. One nugget weighed $45. The excitement was intensified by Nate Vestal at the

> mouth of Buckskin finding ounce diggings in the gravel.
>
> Two cleanups on the Widow this week netted 60 ounces and 8 pennyweights and 88 ounces respectively. Work has been commenced on a ditch to be taken from Alder Creek. This ditch will be two and one-half miles long and greatly increase the production of the Widow. Gen. A. P. Curry washed $21.50 from one pan of dirt selected at random. The total yield of the Webfoot thus far has been something over 71 pounds, or $10,052.65.

A score of other claims reported rich yields that first year. Anything less than an ounce a day per man was considered mediocre or small. In September, Judge William H. Claggett wrote, "Three or four days before leaving home I went over the entire camp where mining has been carried on, and found that less than four acres of bed-rock had been cleaned up, all told. This shows a yield of over $40,000 to the acre. The present output of gold is not less than 1,000 nor more than 1,500 ounces of dust per week."[1] By the middle of August, 2000 locations of placer mining ground, embracing tracts varying from 160 acres to fractions of an acre, were recorded on the books of the Coeur d'Alene Mining District alone, and it was only one of five districts then in existence, the other four being Summit, Beaver, Evolution, and Eagle.

The Old Channel Wash

Parallel with Prichard Creek and on a higher level lies a rich placer area known as the Old Channel Wash. This wash of deep gravel appears to be the ancient bed of a large stream. It is first seen at the mouth of Bear Gulch. From that point it can be traced for 6 or 8 miles along the hills on the north bank of Prichard Creek as far as Eagle Creek where it runs into Daisy and Fancy Gulches. It is an unanswered question whether the wash persists beyond Eagle Creek toward the North Fork. The north side tributary gulches of Prichard Creek all intersect the Old Channel Wash which has been the source of most of the gold found in those gulches. The old channel was discovered by A. F. Parker of Eagle City, or at least the first location therein was made by Parker in October, 1883. By the spring of 1884 the entire area of the wash had been located in claims. Wherever water was convenient and mining could be carried on, the wash gravels proved rich in gold. But a large area, including all from Dream Gulch to Eagle Creek, has never been worked, and remains a vast virgin ground for future development.

The Water Problem

In the foregoing account, I have indicated that one difficulty in mining the old wash was a shortage of water. The same complaint was voiced in the Claggett letter mentioned above and included in Appendix VII. Others beside Claggett interested themselves in the water situation, and we have a number of ingenious and promising schemes developed to bring water upon dry ground or to carry off excess water from the main channel.[2]

The first, in 1885, was a bedrock sluice on Prichard Creek, promoted by a certain Major Hopkins who had the backing of eastern capital. His project was to construct a channel in Prichard Creek down to the bed rock. This channel would be 16 feet wide at the bottom. Sand and gravel would wash down this channel, while boulders would be lifted out by machinery and piled on one side of the ditch. The bottom was to be lined with railroad iron riffles, securely bolted in. Thus, when finished, the bedrock sluice would carry off the main flow of the water and allow the deep gravel to be worked right there on the location, 24 hours a day if desired. Hopkins put 100 men to work clearing a grade for the sluice, but when the grade had been completed from the North Fork only up as far as Eagle, Major Hopkins' eastern backing collapsed and obliged him to stop operations.

Then came a project for a high flume to carry water onto the old wash. A certain Dr. Morgan of St. Louis undertook to build this flume, placing his intake at Paragon Gulch. His flume was to be 6 feet wide and 4 feet deep, and he hoped to supply not only the high placers of the old channel, but also customers in the main gulch with water for quartz mills and other incidental uses. That was a large order for a comparatively small flume, but before a test could be made, Dr. Morgan's enterprise also went bankrupt.

A year later a couple of Californians came into the country and began constructing a ditch line to carry water to the old wash. They made their intake about 4 miles above Murray, and carried their canal over the intersecting gulches in flumes built up on trestles. The flow from the ditch was utilized in Buckskin Gulch where the dirt was all washed from the bed-rock and $10,000 was cleaned up in a few days. At Dry Gulch about $30,000 was washed out by the ditch water. The Californians had demonstrated their scheme to be a success thus far, but they found that their canal was 200 feet too low to be effective in Dream and Missoula Gulches. They sold out to W. J. Hawkins, formerly a banker at Eagle City, who had a fractional claim in the old channel in Dream Gulch. Hawkins had a tunnel built to carry the water to his own holdings. There he cleaned up something like $40,000, taking almost $10,000 in a single day.

Still later, hydraulic efforts were made to cut down the 150 feet solid banks of gravel in the old wash between Dream Gulch and Missoula Gulch. The Coeur d'Alene Mining Company carried the water from Alder and Buckskin Gulches to a reservoir on the divide between Dream and Missoula. Their water supply proved to be inadequate, but the excavated gravel banks high up on the hills near Murray remain as permanent landmarks.

One of the most interesting of all the water schemes was the Nye pump enterprise. The Nye family of Chicago had developed a steam-powered pump which had been used successfully on Lake Michigan at pumping not only water, but sand and gravel from a depth of 30 feet and more. A certain Colonel Ramsey of Minnesota brought the pump to the Coeur d'Alenes, and located it on Prichard Creek near the mouth of Dream Gulch. Just before pumping operations started, however, the machinery was moved by a Mr. Argyle, the agent of the pump company, to a new location in Dream Gulch. All went well until a depth of 20 feet was reached. At this point the water of the side gulch gave out. (They were still above the main creek channel.) The boilers went dry, and there was no water in the shaft. They finally had to secure water from the California Ditch people in order to operate the machinery. Here was a curious paradox of a pump brought in to get rid of water having to buy water. At a depth of 42 feet they found a sloping bedrock, completely dry.

This test was a ghastly failure, and cost Colonel Ramsey $45,000. Adam Aulbach was convinced that the Nye pump had possibilities. Accordingly, with the consent of the Nye people, he relocated the apparatus on the O.K. claim. Here they struck bedrock at 14 feet, with a shaft 12 feet square, and cleaned up 22 ounces of gold.[3] This was scarcely half enough to pay the expenses of pumping. All the boulders larger than a baseball had to be lifted by a windlass. Finally, the whole enterprise was abandoned as too expensive.

In 1890, a Spokane syndicate raised $95,000 to build a watercourse to work the old channel. They laid a 22 inch steel pipe line down Prichard Creek, from the site of Raven down to Dream Gulch, along the creek bed. The question of what to do with their water and the tremendous pressure they had created now faced the company, since they had neglected either to buy or to lease any placer ground. They ran a lateral pipe up Dream Gulch to a great height above the main pipe in the valley, and did some hydraulicking in the old channel. They were soon involved in litigation, and the syndicate went broke before they ever accomplished any profitable operation. The sheriff sold their property for $21,000. It is reasonable to believe that if the big pipe line had been laid along the old channel grade instead of along the creek bed, mining could have been carried on continuously all the way from Raven to Eagle Creek. This failure, therefore, seemed due almost entirely to bad judgment and management.

I have thus described briefly half a dozen rather ambitious attempts to overcome natural obstacles to placer mining on Prichard Creek. None of them was entirely successful, several of them were total failures. The failures depressed the community, but the repeated efforts all testified to the fact that the gold existed there in rich enough quantities to justify a big capital outlay. The fault lay in the planning and execution of the projects.

The Coeur d'Alene Mining Company

In 1900, an important business deal was completed involving Prichard Creek. The Coeur d'Alene Mining Company, under the direction of its manager, Barry N. Hillard, acquired, chiefly by purchase, practically all the placer ground in Prichard Creek and its tributaries, including also the Old Channel Wash. Many of the old claims had two to six owners, so that the deal involved hundreds of persons in all. Under the unified ownership, many lines of constructive work were begun. It was this company who leased their ground to the Yukon Gold company for dredging purposes. And it is this same company which still controls the whole valley and leases ground to individuals and groups.

Eagle Creek Area

Both main forks of Eagle Creek and many tributary gulches have yielded placer gold. Much of the gravel in the lower stretch of Eagle Creek proved to be too deep to work, but many profitable ventures have been made on the upper creek and in the tributary gulches. Rufus Dunlap and Archie Smith operated placers for over 40 years on the upper East Fork, taking out thousands of dollars worth of Gold. Daisy and Fancy Gulches have been especially rich, since they drain the Old Channel Wash. In 1892 Sydenham Mills built a 5 mile flume to bring water into upper Fancy Gulch where he worked for a distance of half a mile. Mills claimed to have taken out $50,000 in gold, some $10,000 of which he surrendered at the point of a rifle one day in the dense timber at the summit of the trail from Fancy Gulch to Murray. This was one of the most spectacular hold-ups in the history of the Coeur d'Alenes.

Trail Creek Area

Trail Creek and its tributary gulches, especially Potosi and Placer Gulches, were extremely rich in placer gold. We have the record of the Myrtle Mining Company for the year 1885. There were five partners who employed 5 or 6 workmen, and in the course of the year in 35 clean-ups, they took out $52,797.90 in gold, including a couple of nuggets worth $300 and $165.[4] Long aqueducts were built to carry water to the diggings, one of them 13 miles in length, from the head of Beaver Creek.

It was in Trail Creek that the largest Nugget of the Coeur d'Alenes was mines, a mass of gold weighing 13 pounds avoirdupois, worth somewhere between $3,000 and $4,000. This huge nugget was the lucky find of Louis Nistler and Peter Borowansky who divided it between them by splitting it with a sharp ax. The nugget achieved considerable notoriety in addition to its size, since it developed that the two finders had been in partnership with a third miner, Clem Reichert, from whom they had kept their discovery secret in order to avoid further division of the prize.

Beaver Creek Area

Beaver Creek was never so rich a field as Prichard, Eagle or Trail. Yet Pony, Trail, California and Prospect Gulches have carried down a great quantity of gold into Beaver Creek, and some of the gravel bars along the creek have proved very rich. It was on one of these bars, near the mouth of Hutchins Gulch, that A. J. Prichard, after years of unlucky speculations, settled down and mined for the final decade of his life. In 1902 and 1903, three dredges were put into operation on Beaver and Trail Creeks. These dredges ploughed up the creek and took out the old town of Delta. Yet it is doubted whether they took out enough gold to pay the expenses.

Output

The question naturally rises, How much gold was produced from the Coeur d'Alene placers? It is a question which cannot be answered with a definite figure of dollars and cents. We can only estimate.

The annual reports of the Director of the Mint show 227,890 fine ounces of gold valued at $4,710,489 as the gold production of Shoshone County from 1884 to 1905. The annual reports of the Idaho State Inspector of Mines show 129,268.33 fine ounces of gold valued at $2,661,019.63 as the county's production from 1906 to 1931. The sum of these figures gives us $7,371,508.63 as the total recorded output. This total makes no distinction between placer and quartz gold, nor between the output of the Clearwater region (before 1904) and that of the Coeur d'Alene district, nor does it account for gold which was produced in recent years as a by-product of the lead and copper ores on the South Side. On the other hand, gold dust sold in the early days in points outside Shoshone County, such as Spokane, would be omitted from the reckoning.

Adam Aulbach in 1897 estimated that $2,500,000 had been taken from the Coeur d'Alene placers since 1884.[5] Another estimate, made in 1902, declares that fully $2,000,000 had been taken from Trail Creek alone, before dredging had begun.[6] The Yukon Gold Company's dredging cleanups from Prichard Creek produced $1,270,000. It seems to me, if I have to venture an estimate myself, that the placer output was not less that $4,000,000 nor more that $5,000,000.

Tables showing the county's gold production year by year are given in Appendix IX.

1. From a letter copied by Onderdonk, p. 130, and included in Appendix VII.
2. Aulbach has reviewed the whole series in a newspaper article, *Scheme to Work Deep Placers on Prichard Creek and Tributaries*, 1932.
3. In an earlier account, *Reminiscences of Prichard Creek, Idaho*, written in 1919, he gives the cleanup as 17 ounces.
4. The itemized record for the year, taken from the *History of North Idaho*, is added in Appendix VIII.
5. Aulbach: *Coeur d'Alene, Idaho's Famous District*.
6. *History of North Idaho*, p. 1061.

Placer Mines, near Delta, Idaho.

Placer Mines, Fancy Gulch, Eagle Creek.

CHAPTER XI

The Quartz Mines

Wherever gold exists in placer deposits it is reasonable to suppose that gold in quartz ledges can be found in the same vicinity. For, after all, the placer gold consists of fragments broken loose from surface outcroppings of quartz ledges and washed down into the gravel of creek beds. Hence, the discovery of a rich placer area always starts a search for the "mother lodes" from which the nuggets and "dust" of the placers came.

In the Coeur d'Alenes, the placers had scarcely been located definitely before the search began for the quartz ledges. This search was rewarded by the discovery of many such quartz veins, both large and small. As early as January, 1884, and all through that pioneer year, prospectors were combing the hills for quartz outcroppings and were staking out their claims all over the North Side. The whole placer area was ribbed and veined with quartz ledges carrying ore varying in richness from a few dollars to hundreds of dollars per ton. However, a circle six miles in radius with Murray as its centre would include all the large and valuable quartz mines.

A couple of placer miners, or even a single miner in a more limited way, can work a placer claim successfully. They can construct their sluice boxes and flumes, do their shoveling and washing of gravel, and make their clean-ups all without the investment of much capital. A quartz mine presents a totally different problem. When a prospector locates a quartz ledge, he generally finds very little ore upon the surface of the ground. He must dig a tunnel into the hillside following the ore vein. He must cut out the gold-bearing quartz and carry it outside the mine. Then he must crush the ore and separate the particles of gold from the worthless rock. A vast amount of labor and of equipment is necessary for a quartz mine to operate upon a large and successful scale. For this reason the development of the quartz mines of the Coeur d'Alenes proceeded much more slowly than the development of the placers. As a compensation to the quartz miner, however, the productivity of his property will outlast that of the placer ground. A placer mine, worked intensively, may last a few years; a quartz mine may last a life-time, or longer.

Outside capital had to be interested and invested before the quartz mines of the Coeur d'Alenes could make much of a showing. Hence, for example, the Mother Lode, a very rich quartz vein, was discovered in January, 1884, but it was not operated until late in 1886.

The first gold quartz mill in the Coeur d'Alenes was built early in 1885 by the Golden Chest Mining Company, a corporation with headquarters in Louisville, Kentucky. The Golden Chest mine is located in Reeder Gulch, a little over a mile above Murray. The mill consisted of five 850-pound stamps and an amalgamating plate.[1] There was no concentrating apparatus in the Golden Chest's first mill, although such machinery was later added. The ore was known as free-milling quartz, and it was run through the stamp mill just as it came from the mine.

This pioneer mine, the Golden Chest, has had a long and interesting history from the discovery of the first ledge of gold-bearing quartz in 1884 (then called the Paymaster) down to the present. Three separate groups of claims were developed: the Idaho group, the Golden Chest group and the Katie and Dora group, all centering about Reeder gulch. Finally all three groups were consolidated into one single corporation, the largest quartz property in the Coeur d'Alenes. In the 1880's and '90's, ownership of the various parts of this enterprise changed hands frequently, operations were intermittent with sudden shut-downs, leases were granted, and the whole period was one of ups and downs.[2] This is characteristic of the quartz mines; their development was not a smooth and peaceful narrative. At one time in the '90's there were 20 stamps falling in the Golden Chest mill and 20 more in the mill of the Idaho company (before the consolidation, of course); hundreds of men were employed and dividends were being distributed to the stockholders. Then, in 1917, after a long period of inactivity, the Golden Chest boomed again, this time not for the gold, but for a rarer and valuable tungsten ore called scheelite (65% tungsten trioxide) which had been discovered years before, but had been a useless product at that time.

The world war had created the demand for tungsten, and the owner of the Golden Chest, Mr. W. Green, received $100,000 (at the rate of $80 for a unit of 20 pounds) for the shipments of scheelite which he sent out by express. For a long time the Golden Chest remained idle again, until December, 1931, when a new organization began to reopen the old workings and to operate the property with new methods and equipment.

The second quartz mine to install machinery was the Golden King, commonly called simply the King, located on the north side of Prichard Creek near Accident Gulch. The King mill which was completed in November, 1886, had ten stamps and three Duncan concentrators, and was thus both larger and more fully equipped that the first Golden Chest mill. Incidentally, the King mine also produced a small quantity of scheelite during the war.

The Mother Lode, one of the earliest and richest quartz discoveries (January, 1884) has, like the Golden Chest, a curious and checkered history. The Mother Lode is located on the south side of Prichard Creek and to the east of Murray—on a steep hillside known as Ophir Mountain where a number of other rich discoveries were made, namely, the Daddy, the Yosemite, the Occident, and the Treasure Box. The earliest efforts to work the Mother Lode employed an arrastre, a primitive machine for crushing the ore by means of large heavy boulders rolled over the quartz on a stone floor. Yet even with this simple equipment, in six weeks of 1887 the owners took out $60,000 worth of gold. In 1890 a five stamp mill with water power was built, and later five more stamps were added to the mill. The mine was an excellent paying property. But it, too, suffered a sudden shut-down followed by years of inactivity and of unproductive leasing operations. Finally, under new ownership, the Mother Lode acquired control of all the other quartz mines of Ophir Mountain, and, again like the Golden Chest, it is the object of renewed interest in 1931 and '32.

A report of 1897 lists the following quartz mills:[3]

Idaho, at the mouth of Reeder Gulch, 20 stamps.
Golden Chest, Reeder Gulch, 20 stamps.
Mother Lode, Ophir Mountain, 10 stamps.
Daddy, Ophir Mountain, 10 stamps.
Yosemite, Ophir Mountain, between Mother Lode and Daddy, 10 stamps.

Joe Dandy, Prichard Creek just below Reeder Gulch, 10 stamps.
Golden King, Accident Gulch, 10 stamps.
Granite, on Granite Creek, 10 stamps.
Frank Jenkins, head of Trail Creek, 10 stamps.
Fay Templeton, Pony Gulch, 5 stamps.
Mountain Lion, west of Murray, 3 stamps.
Buckeye Boy, Dream Gulch, a Dodge pulverizer.[4]

Other accounts tell of the building of a 20 stamp mill in Alder Gulch, two miles north of Murray, which ran only 10 days, being more of a promotion venture than a sound mining enterprise. Also of a 10 stamp mill with steam power near the mouth of Elk Creek, the only quartz gold mill on the South Side, which also ran but a short time. The primitive arrastre was in use at various times on certain other properties.

The above paragraphs sketch very briefly the extent of the quartz gold developments in the Coeur d'Alenes. The years of greatest activity were 1891–1898 when the stamps in 9 mills were falling pretty regularly, and in addition three arrastres and one ball mill were turning. It is difficult to make any estimate of the output of the Coeur d'Alene quartz mines since they were all private enterprises and did not publish much information. The quartz output ran close to \$300,000 per year in those eight big years, and the total production of quartz gold may be set in the neighborhood of \$2,500,000 or \$3,000,000.

1. The stamp mill operates on the principle of the mortar and pestle. The heavy stamps falling upon the ore crush it into minute fragments. Water and mercury are fed into the mill with the ore. The fine particles of gold adhere to the mercury, forming an amalgam which is scraped off the plates and separated in a retort. The stamp mill and the amalgamation process are now obsolete, and the flotation process has largely supplanted them.
2. Adam Aulbach has written a very full and interesting *History of the Golden Chest*, published in the Wallace Miner, January 14, 1932.
3. Adam Aulbach: *The Coeur d'Alene Gold Belt.*
4. The Dodge pulverizer was a fore-runner of the present day ball mill, so universally used in the lead-silver concentrating mills on the South Side. It crushes the ore by revolving it in a metal cylinder along with a number of heavy metal balls.

CHAPTER XII

The Decline of Gold Mining in the Coeur d'Alenes

The picture of the bustling mining activity of the Coeur d'Alenes portrayed in the two previous chapters must contrast vividly with the scene of abandonment and desolation described in the opening chapter. This contrast, indicating a tremendous change in the life of the district, demands an explanation. Why did this melancholy decline take place in a district which started so favorably and proved to be so rich? I have already hinted in the earlier chapters at the causes of the decline, and I shall here elaborate upon those themes.

First of all, we can lay much of the responsibility upon the great discoveries of lead-silver mines on the South Side in 1884 and '85 and the subsequent rush to the new mining field. The very first discoveries of lead-silver ore were on the North Side, however, and as early as August 19, 1883, three lode locations had been filed with the county auditor and recorder.[1] These were the Head Center, Sun Rise and Silver King claims, situated near the headwaters of Beaver Creek, on the north slope of Sunset Peak. The discoverers of this mineralized area were William Sutherland, S. B. Morgan and Charles W. Toole. By April, 1884, a number of lead-silver deposits had been located on the North Side, including the Jack Waite mine.

But beginning with May 2, 1884, the date of the discovery of the Tiger mine in Canyon Creek, by John Carton and Almeda Seymour, the tide of the mining advance into the Coeur d'Alenes began to flow to the South Side. Within that same month of May, 1884, the Polaris, in Polaris Gulch (near Big Creek of the South Fork); the Hunter, above Mullan; and the Black Cloud, on Nine-mile Creek, had all been located, as well as a host of less-known prospects. In August, 1885, came the greatest discovery of all, namely, The Bunker Hill, by Noah S. Kellogg, in Milo Gulch above the present town of Wardner.[2] October 2, 1885, saw the companion mine, the Sullivan, located by Kellogg, O'Rourke, Sullivan and Goetz. The news of these and other lead-silver (galena) discoveries and the subsequent evidence of their great value turned the attention of the mining world again to the Coeur d'Alenes.

The galena rush of 1885 and '86 to the South Fork proved to be an even greater movement than the gold rush of 1883 and '84 to the North Side. William T. Stoll, in his characteristic extravagant style, asserts in *Silver Strike* that half the population of Murray deserted the town after the news of Kellogg's rich strike in Milo Gulch. There can be no question but that many of the North Side miners did join in the rush, hoping at first to locate rich lodes for themselves, later perhaps to get employment in the mines and mills and other enterprises which sprang up in the South Side camps. From 1886 on, the lead-silver activities of the South Side of the Coeur d'Alenes have overshadowed the gold mining enterprise of the North Side both in volume and in public interest. This story of lead-silver mining is a fascinating one, but it lies outside the scope of this study. I can give only this much of it in order to show its relation to the gold story.

A second cause for the decay of gold mining in the Coeur d'Alenes lay within the industry itself. I refer to the weakening of the placer diggings, beginning about 1886. The life of a placer camp is short, for the ground, if worked intensively, is soon cleaned up. The area of easily worked placer ground in the Coeur d'Alenes was small, as has been pointed out before, and in the first three years (1884 to '86) the best of it was worked out and closed. The placer yield began to decline in 1886. It has continued to dwindle year by year ever since, with but few exceptions.

The decline of placer mining is, therefore, not particularly surprising. But the failure of the quartz mines is more difficult to explain. With ledges so large and rich, the quartz mines should have sustained the community for a longer period than a dozen years. There are three factors which seem to account for this unexpected collapse.

The first of these disrupting factors was a change in the nature of the gold ores. The quartz in most of the ledges contained free gold and needed only to be crushed and separated by amalgamation. However, as the ledges were followed lower and farther into the hillsides, the percentage of this free-milling ore decreased, and in its place were found telluride and sulfide compounds. Such a change occurred in the ore bodies of several large Coeur d'Alene mines including the Golden Chest. These compound ores were just as valuable in their gold content as the free-milling quartz, but they could not be treated successfully in the stamp mills or by amalgamation. They required smelting, and in the 1890's it cost $30 a ton to ship ores from Murray to a smelter—that is, to Helena, Montana, or Tacoma, Washington. These ores containing compounds of gold can now be treated very successfully in mills employing the flotation process, but back in the 1880's and '90's the discovery of telluride and sultide ores in a mine was not an occasion for rejoicing. This change of ore, while only a minor consideration in the large problem, undoubtedly did cut into the revenues of some of the mines and did tend to discourage further development.

Another disintegrating factor lay in the ownership and management of the mines. As larger quantities of capital were required to equip and operate the mines, gradually they tended to pass into the control of corporations with headquarters in distant cities. Responsibility was delegated to hired managers. From this situation sprang a long series of incidents displaying ignorance, mismanagement, waste and dishonesty. Stockholders became disgusted and mines were closed because of the failure of absentee ownership and remote control.

The last, but perhaps the most important, and certainly the most interesting factor in the decline of gold mining in the Coeur d'Alenes is a record of "high grading" operations which would do credit to any western thriller of fiction. The term "high grading" describes a system of thievery, the stealing of rich quartz and ores from the mines and the scraping of amalgam from the copper or silver plates in the gold mills. High grading had been the curse of many a mining camp from Colorado to California.

The Coeur d'Alenes in the late 1880's and 1890's became the scene of some high grading on a grand scale.

Adam Aulbach, pioneer newspaper editor of Murray, is the author of this challenging statement, "High grading was the real cause of the failure of gold operations in the Coeur d'Alenes. The gold veins and deposits were not worked out at all. The deliberate stealing that took place left the owners and operators broke." This remark was made in the course of a conversation on July 30, 1931, and was followed by the citing of a number of instances. For several months this topic remained as an intriguing theory in the mind, and hypothesis hardly definite enough to write into a scholarly paper. But in January and February, 1932, Mr. Aulbach gathered up his evidences and recollections into a series of three newspaper articles published in the Wallace Miner, "High Grading Thrived With Development of Quartz Gold Mines in Coeur d'Alene," in which he made out a very positive case in favor of his premise. His conclusion, equally as forceful as the statement at the beginning of this paragraph, reads as follows: ". . . With one exception, namely the Golden Reward, which never had an adequate ore showing. . . not a single gold quartz operating property was ever closed down for the want of sufficient mill feed."

Among the numerous incidents related by Mr. Aulbach in support of his conclusion, three stand out as especially significant. One of them tells how, after a rich cleanup at the Golden Chest, a large gold brick valued between $4,000 and $5,000 was left by the sub-manager in the office of the mine before being carried to Murray for safe keeping. The sub-manager stepped out of the office for less than five minutes, and when he returned the brick had disappeared. Only two workmen were employed in the nearby mill at the time, and these two along with the sub-manager were the only persons who could have seen or touched the brick. Yet, in spite of repeated searches by the company and investigations by the sheriff, no trace of the stolen brick was ever found. This theft was one of the contributing causes of the final shut-down of the Golden Chest mine, which was the hardest blow the North Side gold country ever suffered.

Another sensational incident of high grading occurred at the Pilot mine in Gold Run Gulch. A rich strike had been made and about 30 sacks of ore had been stored in the blacksmith shop of the mine. At this point the mine was sold to a prominent Wallace man who paid $20,000 in cash on a $100,000 option. The new owner increased the crew and continued the digging, Then one day ten sacks containing the richest ore disappeared from the blacksmith shop. They had apparently evaporated into thin air, for no sign of them ever appeared again. The Wallace owner dropped the purchase and took his loss of about $8,000 worth of ore in addition to his cash payment. Years later, one of the high graders confessed on his death bed, revealing how 14 out of the crew of 18 were involved in the theft, how they packed the ten sacks out over the Old Channel Wash and thence to Murray, and how they crushed the ore in hand mortars and washed out the gold in pans.

The third incident occurred at the Mother Lode, and led directly to the closing of that mine. The owner of the Mother Lode caught a stranger gathering rich quartz specimens from his ore dump and putting them into a sack. The mine owner warned the volunteer prospector to get off the dump and leave his ore specimens behind. When the latter refused to go, the owner had him arrested and tried before a justice of the peace at Murray. The accused demanded a jury trial and then made so specious a plea before the jury that a verdict of "Not guilty" was returned. The mine owner, finding that he could not protect his property in the courts, closed down both mine and mill and left the country.

Yellow gold exerts a powerful temptation upon unscrupulous men, and a mine owner or operator must select carefully the workers he employs. The failure of the Coeur d'Alene mine owners and operators to stamp out high grading undoubtedly explains, at least in part, why all the quartz mines and mills had ceased operating by 1900.

1. This date was given out by Harry A. Rogers, auditor and recorder of Shoshone County, in a newspaper interview, 1925. Dates of other lead-silver discoveries are taken from the *History of North Idaho*.
2. The location date for the Bunker Hill is September 10, 1885.

Golden Chest Mine and Mill.

CHAPTER XIIi

The Yukon Dredge

The most spectacular development in the gold belt of the Coeur d'Alenes in recent years was the dredging operations of the Yukon Gold Company in Prichard Creek. The idea of dredging the creek channel to recover the deep placer gold seems to have originated about 1910. The Yukon Gold Company, a New York corporation dominated by the powerful Guggenheim interests, set up and office in Murray, and proceeded to prospect Prichard Creek channel from Granite Gulch to the North Fork. Satisfied that the gold values would justify the expense of digging, the company leased the channel from the Coeur d'Alene Mining Company and commenced its operations.

In the fall of 1917 the great dredge was assembled in a deep pit on Prichard Creek near the mouth of Dream Gulch. Most of the machinery came in from Alaska where the company had operated dredges on a large scale. The boat measured 106 feet in length, by 44 feet in width, and it drew 8 feet of water. Actually it took 9 feet of water to float the dredge in operation and 12 feet were preferred for safety.[1] The dredging machinery was run by 6 electric motors aggregating 400 horse power. The electric power was supplied by the Washington Water Power Company of Spokane, The dredge received the power at 6,600 volts and transformed it down to 440 volts aboard.

As soon as the machinery had been installed, and without the loss of a single day, the dredge started operating. On December 5, 1917, the endless chain of 76 huge steel buckets began digging into the gravel and boulders of Prichard Creek. The work continued for 8½ years almost uninterrupted, the final clean-up taking place on May 29, 1926. To show the magnitude of the dredging operations, here are a few interesting facts.

The dredge worked 24 hours a day, employing about a dozen men in 3 shifts of 8 hours, and it never closed down except for bad weather, low or high water, or breakdown of machinery. In one year they worked 7,345 hours out of a possible 8,760. They averaged over 200 cubic yards of gravel per hour, with 2 tons to the cubic yard. In the 8½ years, the dredge worked over 11,000,000 cubic yards. Mr. Stewart Campbell, state inspector of mines, is author of the statement that the dredge handled "more tons per day than all the mines in the Coeur d'Alenes."

The Yukon Gold Company occupied the old court house as its office in Murray, used the old county jail as a store room, and set up its retort house in a small building back of the jail. Mr. D. H. Ferry was manager of the operations the entire period. In addition to the payroll of the dredge workers, which ran above $25,000 a year, the company spent on the average $2,500 a year for the clearing of the ground in advance of the dredge, and nearly $10,000 a year for supplies and office equipment. Most of this money was spent in Murray, and it gave that waning town a final big boom of prosperity.

Beginning in December, 1917, the dredge moved down Prichard Creek on the south side of the valley, and turned around at the King property (Accident Gulch) in 1920. The ground below the King proved to be too deep to be worked profitably. The dredge then proceeded back upstream, reaching Murray early in 1922, and spending the rest of that year working over the old townsite. One whole street (Gold Street) was taken out, and the dredge advanced close to Main Street at several points. It ran on up the creek to the site of Raven, just below the mouth of Granite Creek. At that point they found a solid bar of shallow bedrock blocking the channel completely; it was too shallow to float over, and too wide to cut through, profitably. Besides, they had almost reached the upper limits of the gold-bearing sands. The last clean-up was made May 29, 1926, and the work of dismantling the machinery began immediately. The motors were sold to the Coeur d'Alene Hardware and Foundry Company of Wallace, and most of the other valuable equipment to a dredging company on the Rogue River, Oregon. The crew was paid off on July 1, and the abandoned and dismantled hull of the dredge was left stranded in its last pool, where it may still be seen by all who pass on the nearby county road.

In the course of 8½ years the Yukon Gold Company dredged about 7 miles of Prichard Creek, cleaning from 21 to 42 acres of bedrock each year, depending upon the depth of the gravel and other operating conditions. Below Murray the bedrock lay on the average 30 feet below the surface; above Murray, about 25 feet down. In addition to digging away the gravel and boulders which lay above the bedrock, the dredge dug 2 to 4 feet of the bedrock itself in order to make sure of saving the gold. In the process of dredging, the bottom of the creek was upheaved and turned up-side down, leaving a solid pile of huge boulders over the surface of the valley, a feature of the landscape which now distinguishes Prichard Creek from all other valleys of the Coeur d'Alenes. In addition, at intervals all along the creek parallel with the boulder piles may be found the rusty steel buckets and other parts of machinery which were broken and cast away in the course of the dredging.

In the course of the 8½ years the Yukon Gold Company recovered $1,270,000 worth of gold from the creek bottom. In 1918 the big production by the dredge gave Shoshone County first rank among the counties of Idaho in gold output—that honor for the first time since the pioneer gold strike of 1861 (the Clearwater placers). This great quantity of gold was washed out of the gravel in sluices aboard the dredge; the metal was amalgamated with mercury, and separated by retorting; the gold was finally molded into bars and carried to Wallace by stage for shipment to New York. From the total output, the Yukon Gold Company took a net profit of about $500,000 without deducting for depreciation.

1. This fact and many others which appear in the course of this chapter were learned in conversation August 18, 1932, with Mr. Howard W. Staples of Moscow, Idaho, who was assistant manager on the dredge for 4½ years.

Abandoned Hull of the Yukon Dredge,
in Prichard Creek.

Mass of Boulders Left by the Yukon Dredge
in Prichard Creek, below Murray, Idaho.

CHAPTER XIV

Gold Mining in the Coeur d'Alenes Today. The Prospect

Ever since 1926, the last year of the Yukon dredging operations, the gold production of Shoshone County has been extremely low. The figures are less than $15,000 a year on the average—lower than ever before in the history of the district. It would seem as though gold mining in the Coeur d'Alenes had reached the final fade-out leading to oblivion.

There has never occurred a complete cessation of gold mining in the valleys of the North Side. Every year in the early spring months, the casual motorist as he drives from Wallace to Murray can see placer digging in Trail Creek. And if an observer would look into the side gulches of Prichard Creek, such as Buckskin, Nugget and Dream Gulches, he might see the same picture of placer activity year after year. But this working area is small and the working period is short, since these small gulches run dry early in the summer. And the general impression which the present-day visitor to the North Side gets, when he sees the stark line of dredge tailings in Prichard Creek and the deserted and ruined old stamp mills up the valley and the dilapidated sidewalks and empty houses of Murray—his impression is that the golden days are past and gone and all that remains is a memory.

However, the story is not so easily told and the question so lightly dismissed. The nation-wide economic depression which began in the fall of 1929 has altered the situation. The mining industry in the Coeur d'Alenes, along with many industries in many communities, has suffered heavy losses. Prices of lead, silver, and zinc have fallen to new low levels in the history of the district. Many mines have closed down entirely; others operate upon greatly reduced time schedules. Miners are out of work; store-keepers are losing trade; in short, hard times are here in the Coeur d'Alenes. But notice how this deflation effects the gold industry. Since gold is the standard money metal whose price never varies ($20.67 the ounce of fine gold), and since the values of other commodities are expressed in terms of money and hence of gold, then as lead and silver and zinc decline in value (as expressed in price), gold increases in value proportionately. Thus, as lead, silver and zinc become less and less profitable to mine, in the face of falling prices, gold becomes more and more profitable to mine.

As a result of this sensational reversal in the metal markets of the world, as well as of the general decline in employment and business in other industries, the attention of the country has been directed toward gold mining as it had not been for a decade or longer. Groups of amateur prospectors, recruited from the ranks of the unemployed, have swarmed into the hills and valleys of the western states, seeking "pay dirt" and hoping to make a lucky strike. Some of the argonauts of 1932 have entered the Coeur d'Alenes, most of them to meet early disappointment and disillusionment, but a few of them to settle down to the hard and tedious work of placer mining which may reward them with golden dust in their pans.

But more serious than these amateur and somewhat hap-hazard efforts is the attempt to interest capital in reopening the quartz mines of the early days. Extensive work along this direction has already been started at the Golden Chest mine by a group of Kellogg enterprisers under the leadership fo Mr. Charles E. Moore. The old tunnels and stopes of the Golden Chest are being opened again and arrangements have been made to treat the ores with the modern technique of the flotation process. If the Golden Chest venture proves successful then similar attempts may be made at the Mother Lode and other promising ledges.

Thus gold mining really hangs in the balance today in the Coeur d'Alenes. There is the prospect of a revival which may resemble the old pioneer days in activity and output. There is the equally possible prospect of further stagnation and decline in the event of failure or of only mediocre success in the present operations. The few old-timers of the '80's who still live and work around Murray have retained their hope in the future, and their confidence in, the resources of the district. Adam Aulbach, with pride and conviction in his voice, sums up their optimism when he declares, "They'll be digging ore here when the South Side is finished."

Perhaps they will. Who knows?

APPENDIX I

Pioneer Gone

Andrew J. Prichard, the discoverer of gold in the Coeur d'Alenes, died at Murray last Friday night (October 3, 1902) after a long illness.

Mr. Prichard was a native of Connecticut, 72 years of age, and a veteran of the Civil War. After years of prospecting through the West, during which he passed through the Coeur d'Alenes once and located a claim now known as the Evolution, near Osburn, and in 1882 coming back to prospect it further, he went across the mountain and discovered placer gold. A year later the rush to Eagle began and the Coeur d'Alenes have since been known throughout the mining world.

Like most discoverers, Mr. Prichard received but a small amount of the wealth which he was instrumental in adding to the world's store, leaving his wife and four children in but moderate circumstances.

For several years he has been failing in health, owing to age more than disease, and his death has been expected at any time for two or three months.

He was buried Sunday at Murray, under the auspices of the G. A. R.

Wallace Press, October 11, 1902.

APPENDIX II

Prichard's Story

I came to the Coeur d'Alene country by way of Montana from New Mexico to Colorado in the fall of 1878, and had never heard of Cannon, Warner & Company before my arrival at the post. I found the quartz lead, known as the Evolution lead, on my way in the month of November, 1878. On my way I left the road and followed the ridges, crossing small tributaries looking for quartz, and one night did not get into camp until midnight. It being late in the fall I went to Post Coeur d'Alene, thence over to Heyden's (Hayden) Lake, where I did some work for Heyden, and remained until January. While there, hearing that Cannon, Warner & Company, wanted some logs, I went to Spokane Falls to see them and took a contract which was to have been completed by July 1, 1879, that being about the time the water runs down so as to enable one to prospect, but early in the spring the unusual high water swept away Cannon & Company's boom, consequently they could not receive the logs. The money I brought to the country being invested in the contract, Cannon & Company sent for me and I went to the Falls, where they made further arrangements with me, increasing the contract so as to give me employment until they could replace the boom, which they failed to complete until late in the fall. My first acquaintance with Tom Irwin was in January, '79, while at the Falls. Finding him gentlemanly and an old prospector, we talked over the prospects of the country, in which I spoke of my finding what is now known as the Evolution lead. In the spring while I was at work Mr. Irwin came up on a prospecting trip on the South Coeur d'Alene. The water being high, he left some of his packs at the mission and went as far as Nigger prairie with some men going to Montana, one of whom had worked for me. They had to take the high-water trail, and its being blocked with fallen timber and brush, caused much delay. The men going to Montana, being short of provisions, ate Irwin out, so he had to return, doing very little prospecting. On his return he found the lead near Miners' Camp, to which he afterward returned and put up his cabin in the summer of '79. As the editor claims, with much emphasis, that I did not come before the fall of '80, I will add that I have a bill of goods bought of C. F. Yeaton, post trader of the Coeur d'Alene, lying before me dated July 7, 1879, and one of August 1, 1879, amounting up to that date to $610.26. I came in the fall before I put up my cabin at Evolution, the fall of '79, above Miners' Camp and worked my lead through the winter, doing very little execution, having no tools to work with and no money to buy, having only a silver dollar left on arriving, and in February I commenced prospecting a small gulch called Prospect Gulch, putting in sluice boxes in March, and to my knowledge, taking out the first placer gold on the Coeur d'Alene. It being a small gulch, the water did not last long, but I got some very nice specimens of coarse gold.

History of North Idaho, p. 985.

APPENDIX III

Evolution,
Jan. 8, 1883.

Friend Burk, Dear Sir:

I will drop you a few lines as I have something to write hoping they may find you O.K. Wishing to assist my infidel friends whenever opportunity offers, and to assist in throwing into our ranks all I can, I give you this information and hope you will assist in extending it to others through the Leagues, and if possible not outside of our brethren and sisters. I have discovered a large section of gold country, have found gold on five streams, but have only prospected two of them, enough to know they will pay.

One of them is between 20 and 30 miles long, the other 12 to 20, with average width of 60 or 70 rods with gold in paying quantity the entire length. Also with the tributaries which will undoubtedly pay. The gold is coarse and of good quality with bed rock from five to 12 feet. As near as I can judge from what I have done there is enough to give employment to 15 or 20 thousand men, and I should like to see the Liberals get the Lion's share, if possible.

I have spent three years of the hardest work and worst exposure of my life working it up by myself, with no assistance or encouragement from anyone and now I intend to reap my harvest, and should like to see my friends enjoy some of it. If the Liberals will act promptly and not get excited and work together they can get a good slice of it and also build a city that they can control and have good laws and regulations. The country is literally traversed with ledges of quartz, and plenty of good timber and water. As I am virtually shut up in winter 50 miles from post office, I can do but little to give our friends the information and I know from what I learn that it can't be kept long and before next summer closes there will be as big a rush as has been witnessed for years, and if you will get information to as many Leagues as possible without letting the public know, you will lose nothing by it, and you may do hundreds a great favor.

Perhaps it might be a good plan to write to some of the men in the lecture field and get their assistance, for instance, Follett of Illinois, Green of New York, and others. Use your own judgment and let as many good Liberals know as you can and come yourself in the spring if you can. If you can't send me a power of attorney and I will locate mines and have them worked for you on the waters of the Coeur d'Alene river, in Coeur d'Alene mining district and it shall cost you nothing, and I will divide profits with you. If D-S. is there and you can trust him to keep it, tell him. The new mines are about 40 miles from my place on the road back in the mountains, which is a poor trail, as I have only cut it out sufficient to get through and pretty rough at that. Men don't want to come without provisions, for it will be 90 miles over bad roads to get supplies. Can't get in before the middle of May from that side, and only a week or two earlier from this.

I have given enough that you may be able to have an understanding and don't be afraid to work. It's no humbug. Don't know how soon I can get this to the office, as I have to depend on the repairs men on the government telegraph line which runs past here. They have a station three miles below here, they being our only neighbors.

Now I will give directions so that by inquiry you can find me. I am on the old Mullan Road, 25 miles west of the Bitter Root mountains, about 125 miles from Missoula and 50 miles east of Fort Coeur d'Alene, Idaho, which is the nearest place to get supplies, and 23 miles east of the Old Mission. If there are any coming from the Eastern states their best route will be via California, Portland, on the route of the N. P. R. R. to Rathdrum, which is 12 miles from Coeur d'Alene, where they will have to fit out as they can get nothing nearer.

Hoping to see you in the spring, I remain

Yours,
A. J. PRICHARD,
Coeur d'Alene, Idaho

Original letter in possession of Jerome J. Day, Wallace, Idaho.

Evolution,
Jan. 7, 1883.

Mr. C—:

Dear Sir—As I came across your name in the Truthseeker, I take the liberty of addressing you a few lines in hopes of giving my "liberal friends" a little assistance. I can not enter into details in a short letter, but will give enough to give you an understanding of what I wish and will ask you as a gentleman and brother liberal not to make it known outside the Liberal League and its members.

I have made a discovery of a gold-bearing country that will give employment to at least 15,000 to 20,000 men. There are two streams that I have prospected well; one is sixteen to twenty-five miles long, as near as I can judge; the other twelve to sixteen miles and an average width of sixty to seventy rods; have found gold on three other streams of near the same size, but have not tested them enough to know how they will pay. The two streams I speak of will pay their whole length and probably the most of their tributaries, with an abundance of good timber and water. Bedrock from five to twelve feet. Gold coarse and of good quality. There are two good and natural town sites where will be built cities representing thousands in less than two years, and the country is traversed with hundreds of mineral bearing lodes of quartz. And now for good reasons which I have not time to explain I would like to see as much of this go into the hands of the liberals as possible, and also see them build a city where they can have their own laws and enough of this vast mining region to support it, which they can do if they will go at it cool and work together. I have spent four years here by myself looking and working it up. I first discovered and located a lode on the Mullan road, and not having much means to open it up, I spent all of my spare time looking for placers, not anticipating finding extensive mines, only something to help me open my lode, but I have found a richer and bigger section than I supposed lay undiscovered in the Rocky range, and now if you will convey the purport of what I give you to as many leagues as you can on this coast, and request them to get together and keep this information to themselves, they can secure the "lion's share." I am in the mountains, fifty miles from a postoffice, and can do but little in winter, for the snow gets from three to four feet deep here. I will give directions how to get there and what is needed. My location I call "Evolution" as that is the name of my lode. It is on the old Mullan road to Montana, I am fifty miles east of Fort Coeur d'Alene and twenty-three miles east of the old mission. The Northern Pacific railroad runs within twelve miles of the post, where there is a town called Rathdrum. Parties coming will want pack animals, as the new mines are back from my place on the road forty miles in the mountains, with but poor pack trails yet, as I have not had time to cut them out more than enough to get through, and they will want supplies for a month or two, as there will be no chance of getting anything after leaving the post at present. Probably the best place for those that have to buy horses would be to stop at Spokane Falls, Washington Territory, which is thirty miles from the post, and perhaps they might do better in provisions and groceries, tools, etc.

Now, if there are many that conclude to come they might leave the impression along the road that they were going to Montana and give as many liberals as possible a chance to get in before they get up and excitement.

I shall be down about the middle of April and that is as soon as they need to get to the post, as they can't get into the mountains until the water runs down, and sometimes it keeps up until the middle of May, so they need not be in a hurry getting an early start. There are but three or four that have a knowledge of it (the discovery of gold) here, and that is merely a suspicion. They have agreed to keep it. I am a little afraid of one of them or I would have waited a little longer, but shall try to get ready. I wanted to prospect a little more before a rush commenced, for it will surely come. Now, hoping that my friends my profit by the knowledge I have given them, I remain,

Yours fraternally,
A. J. Prichard.

Letter purported to have been written by A. J. Prichard. Published by the Spokane Review in 1888. History of North Idaho, pp. 987–988.

APPENDIX V

Circular

Northern Pacific Railroad Co.,
General Passenger Department.
Circular No. 6.

St. Paul, Minn., February 1, 1884.

The wonderful rich mineral wealth of the Coeur d'Alene mountains and the tributaries of the river of that name, has been heralded to the world, and old prospectors and miners who are conversant with the history of the banner districts of Nevada, Colorado, Utah, Arizona and New Mexico stand amazed at these new fields, which are

UNEQUALED IN RICHNESS AND EXTENT,

the yield being practically inexhaustible, rendering impossible any over crowding of the district by reason of too great an influx of prospectors and miners. *There is more than enough for all who come.* The minerals consist of silver, tellurium and free milling gold.

$100 PER MAN PER DAY

are being taken out of the rim rock of the gulches, while in gulches $25 to $40 per man per day are being panned out. The claims are all very rich, and are located in the gulches of the North Fork of the Coeur d'Alene river, Eagle, Prichard and Beaver Creeks, streams running into the Coeur d'Alene river. The Prichard is the chief stream, flows two thousand inches of water, and it is said, does not freeze up, the altitude being only 2,400 feet.

$200 NUGGETS OF GOLD.

Rich placer deposits have already been discovered for a considerable distance on Prichard creek, and the same distance on Eagle creek, the creeks being known by the latter name from the point where they come together. Even where few developments have been made, the prospects obtained show beyond a doubt that nature has been very lavish with her favors, as from 12½ to 50 cents per pan is regularly obtained, and in some instances as high as $1.75 per pan. Wherever the bedrock has been uncovered, beautiful, rich dust has been scooped up by the lucky owners. On Prichard creek can be seen valuable placer claims that have been opened, and not in a single instance has the work of development been advanced without a correspondingly rich return in bright, golden nuggets, some of which have been sold for $50, $100, $166, and $200.

FREE GOLD QUARTZ.

An intense excitement has sprung up in regard to the quartz deposits of this district, the immediate occasion of this being a "find" of a valuable gold quartz lode at the head of Prichard creek. The vein has been traced on the surface for a distance of five hundred feet, and the outcroppings are very prominent. The ore taken from the vein shows a great amount of free gold; in fact it fairly glistens with it, plainly discernible to the naked eye. Very little attention has been paid to quartz heretofore, but this strike has done much toward stimulating prospecting in this line. A gold vein near Miner's Cabin, on which a great deal of development work has been done, is from 18 to 24 inches wide, and hardly a chunk can be picked up on the dump and broken which does not show free gold. A shaft has been driven down on the ore body to the depth of 80 feet, and in the bottom the vein is strong 20 inches wide. A tunnel advanced 350 feet would tap the ore body at nearly 500 feet. This ore assays from $28 to $196 per ton gold, the former being the smallest assay ever obtained, as high as $900 in gold per ton being not infrequent, while certain specimens have assayed $10,000 per ton gold.

A MAMMOTH GALENA BELT,

the most extensive known at the present day, is being developed on Beaver creek. The vein can be readily traced on the surface for five or six miles, the ore carrying from 80 to 90 ounces of silver, and 35 to 40 per cent lead per ton. At many points the veins have been stripped for hundreds of feet, and everywhere show a remarkably high percentage, not only in silver and lead but up in the hundreds in gold. The outcrop frequently measures 60 feet in width. Such is a brief sketch of the Coeur d'Alene mines, which surpass in richness and volume the most fabulous quartz and placers ever discovered, even the famous deposits of "Potosi" being inferior to those which underlie the mountains of the Coeur d'Alenes. As the mines of the of the old world, some of which have been worked ever since the eleventh century, are still employing thousands of men, the conclusion to be drawn in regard to the Coeur d'Alenes, a region far superior in every way, is that they are inexhaustible, and although thousands may work them, there will still be for thousands more, "Millions in it."

Respectfully,
Chas. S. Fee,
General Passenger Agent.

The Coeur d'Alene Nugget. Vol. I, No. 1. Saturday, March 15, 1884.

(We publish the above circular free of charge, omitting nothing but the routes and rates. It is a bigger story than we can tell, but we shall be glad to find that it is not a roseate view. We hope Mr. Fee knows whereof he writes.)

APPENDIX VI

Coeur d'Alene Mining Laws

(Following is a copy of the by-laws adopted by the first miners locating in the new regions:

Local laws of the Coeur d'Alene mining district located on Prichard and Eagle creeks, Shoshone county, Idaho Territory:)

BY LAWS

Sec. 1. All locations on lodes or veins of quartz to conform with the United States laws of May 10, 1872, as nearly as practicable, viz: Fifteen hundred feet (1500) in length, by six hundred feet in width.

Sec. 2. Placer mining claimants shall be allowed twenty acres, to be located so that the length shall not exceed eighty rods.

Sec. 3. Each location shall be represented by the locator or his authorized agent in locating and recording.

Sec. 4. No person shall be restricted to one claim, but may locate one claim in any stream or gulch where vacant ground may be found. But no person shall be allowed to locate more than one claim on the same stream or gulch. Persons shall not be prohibited from holding claims acquired by purchase.

Sec. 5. Claimants shall have one year from the first of January succeeding the date of location to work their first annual assessment which shall be one hundred dollars. Each year thereafter claims shall be represented by twenty dollars worth of labor each month after the first of June until the first of November after the first year's assessment. Furthermore, all claims shall be considered laid over from the first of November to the first of June. All necessary work, such as making roads or trails, building houses, or any improvements in opening or working a claim, will be allowed five dollars per day as assessment labor.

Sec. 6. Claimants will be required to record their claims in the district record within fifteen days from the date of location.

Sec. 7. The oldest or first claimants shall have the first privilege of water, but shall not prohibit others from using the surplus water, and all claimants shall be required to return the water to the channel of the stream for the benefit of those below.

Sec. 8. Several miners may form a company for the purpose of opening and working mines in placer claims, when such claims are contiguous, and the labor performed by said company shall represent their several claims, although the amount of labor for the convenience of opening and working may be done on one claim.

Sec. 9. Difficulties arising between parties in the mining district shall be settled by arbitration, each disputant to be allowed an equal number of arbitrators, and in case of a tie on decision said arbitrators will have power to call an assistant.

Sec. 10. All claims located prior to the date of the adoption of these by-laws shall be respected just the same as those made after said date.

Sec. 11. The records of the Coeur d'Alene Mining District, in Shoshone county, Idaho Territory, shall be kept at the house of A. J. Prichard, Recorder, near the confluence of the Eagle and Prichard creeks.

Sec. 12. On the written application of 12 or more miners, the Chairman shall cause three notices to be posted up in three conspicuous places, giving the days' notice of a meeting, said notices to specify the object and business to be transacted at such meeting. To make any changes in the present by-laws between the first of November and the first of June the following year shall be illegal.

Sec. 13. The laws shall take effect from this date, and any laws or regulations previously enacted that conflict with these laws shall be considered repealed.

The Coeur d'Alene Nugget, Vol. I, No. 5, Wednesday, April 9, 1884.

APPENDIX VII

A Review of the Placer Mines

"Before the claims on these tributaries could be opened for work, the water supply from the melting of the snow failed, and the high water in the main gulch discouraged, if it did not wholly prevent, any systematic attempt to open the claims. Then, just when everybody was getting ready to work, the entire camp found itself involved in litigation. Suits were begun in the district court to recover possession of nearly every claim in the camp. This litigation absolutely paralyzed development down to the 6th day of August, when the court adjourned. The result of the trials was to leave the claims in the hands of those who were called the 'jumpers,' and for the purposes of development, the 6th day of August last was the beginning of the life of the camp. Everything of any consequence in the shape of placer mining has been done since that date, which makes the camp at this time, to all practical intents and purposes, less than two months old. There being no water in the tributaries with which to mine, the development has been confined to the main gulch, commencing about two miles above Eagle and running up the gulch a distance of about seven miles. The gulch itself is more like a mountain basin than anything else. It is from 40 to 120 rods wide, and on the north side, having a general parallel course with it, there are several old river-beds or channels having a vertical elevation above the gulch of from 75 to 300 feet, through which intersecting ravines have cut, and the main gulch is mostly supplied with gold washed down from these old river-beds. This 'old wash,' as it is commonly called, has been traced for a distance of thirty miles. Thousands of acres of it have been located. In the absence of water to work it, it has thus far been only slightly prospected by shafts and tunnels, the miners waiting for some one with sufficient capital to bring water upon the hills.

"Messrs. Walsh & Hubbel, old and experienced placer miners from California, are now digging a ditch for this purpose, and when this and other ditches are completed, the Coeur d'Alene region will be an eye-opener to those who imagine that there is nothing there. Men are now packing gravel in sacks from the sides of the mountain upon their backs to water, and making from $25 to $50 per week. When a sufficient number of mining ditches shall be dug, the whole aspect of things will be changed. The water being taken from the main gulch will enable the gulch miners to work to better advantage. Troubled as they are with a superfluity of water, hydraulicking the hill ground will make available the immense resources of wealth which it contains, while the muddy water, turned back into the main gulch, will puddle the bed of Prichard Creek, so that the gulch claims can be successfully opened in the deep ground. Three or four days before leaving home I went over the entire camp where mining has been carried on, and found that less than four acres of bedrock had been cleaned up, all told. This shows a yield of over $40,000 to the acre. The present output of gold is not less than 1,000 nor more than 1,500 ounces of dust per week. Several of the claim-owners ship their dust directly to Helena, and my estimate is based upon my knowledge as to what these claims yield, the quantity of dust purchased weekly by the banks, and taken in by a few of the merchants. It is a mistake to suppose that Prichard Gulch comprises the entire gold-field. On Trail Gulch several claims are opened, and are proved to be rich, while there is as much to show that Beaver Creek is rich as Prichard Gulch itself last spring. There are two great wants to be supplied: first, a number of mining ditches to carry water upon the old wash; and second, a practicable wagon-road into the mines, over which mining machinery can be hauled and coaches travel."

Written by Judge William H. Claggett, from Murray, September 22, 1884, as quoted in Onderdonk, p. 130.

APPENDIX VIII

Record of Myrtle Mining Company, 1885

Pat Mahoney, James Osborn, Mike Connors, Dick Buckingham, Dave Alvard.

These men, with 5 or 6 employees, worked with wheelbarrows on a claim a mile above Delta (on Trail Creek).

Cleanups for 1885.

March	24–28	$ 495.00
March	30–April 4	1621.50
April	7–11	1054.50
	14–18	689.50
	21–25	1158.00
April	27–May 2	1188.00
May	9	1010.00
	11–16	1690.50
	19–23	1700.15
	26–30	905.50
June	2–6	585.50
	8–13	865.25
	18–20	935.00
	22–27	2595.00
June	30–July 4	2282.00
July	6–11	2040.00
	13–18	1916.50
	21–25	803.00
July	27–August 1	935.00

August	3–4	168.00
August	29–September 4	553.50
September	5–12	561.00
	14–20	2878.00
	21–27	3891.00
September	28–October 4	4131.50
October	5–11	2167.00
	12–18	1385.00
	19–25	1826.50
October	26–November 1	2814.00
November	3–8 (including $165.00 nugget)	1896.50
November	9–15	1638.50
	16–22	1030.00
	24–29	1549.50
December	1–6	1131.00
	7–11	707.00
	Total	$52,797.90

APPENDIX IX-A

Production of Gold, Shoshone County

Year	Fine Ounces	Value
1884	12,500	258,375
1885	18,220	376,607
1886	8,823	182,371
1887	7,367	152,276
1888	10,250	211,867
1889	8,433	174,310
1890	8,000	165,360
1891	10,000	206,700
1892	11,000	227,370
1893	14,748	304,841
1894	17,531	362,365
1895	18,439	381,134
1896	17,369	359,017
1897	16,404	339,070
1898	13,011	268,937
1899	8,602	177,803
1900	5,754	118,935
1901	4,915	101,593
1902	4,761	98,410
1903	7,651	158,146
1904	2,226	46,015
1905	1,886	38,987
	227,890	$4,710,489

The gold there accounted for is derived entirely from the placers and gold-quartz veins near Murray, the lead-silver ores containing so little gold that the quantity is negligible.

From reports of the Director of the Mint. Ransome and Calkins, p. 82.

APPENDIX IX-B

Production of Gold, Shoshone County

Year	Fine Ounces	Value
1906	3,244.32	$ 67,060.00
1907	3,435.	71,001.45
1908	4,105.	84,850.35
1909	7,155.55	147,905.30
1910	3,110.	64,283.70
1911	3,054.37	63,133.83
1912	4,446.6	91,911.00
1913	3,510.	72,551.70
1914	4,052.	83,754.84
1915	4,500.	93,015.00
1916	2,600.	52,000.00
1917	4,753.	95,060.00
1918	14,628.	302,360.76
1919	10,191.	210,647.97
1920	9,000.	180,000.00
1921	7,217.33	149,182.21
1922	6,049.35	125,040.06
1923	11,873.17	245,418.42
1924	8,026.31	165,903.83
1925	6,259.65	129,386.97

1926	5,060.43	104,599.09
1927	436.04	9,012.95
1928	440.57	9,106.58
1929	637.26	13,172.16
1930	810.66	16,756.34
1931	672.72	13,905.12
	129,268.33	$2,661,019.63
For the years 1884–1906	227,890.	4,710,489.00
	357,158.33	$7,371,508.63

From the annual reports of the State Inspector of Mines.

CRITICAL BIBLIOGRAPHY

I. Books.

Brosnan, Cornelius James
History of the State of Idaho. Illustrated.
New York, Scribner, 1918.

A brief study valuable for accurate dates of early settlements.

Cody, Rev. Edmund R.
History of the Coeur d'Alene Mission of the Sacred Heart. Illustrated.
Caldwell, Idaho, The Caxton Printers Limited, 1930.

An interesting account of the first white establishment in the Coeur d'Alene region.

Durham, N.W.
History of Spokane and Spokane Country, Washington, From Its Earliest Settlement to the Present Time. 3 vols. (1 vol. history, 2 vol. biography) Illustrated.
Spokane, Chicago, Philadelphia, the S. J. Clarke Publishing Co., 1912.

Vol. I, chapters 38 and 39, deal with the discovery and development of the Coeur d'Alene district. The material is good, but brief, and evidently has been drawn from the Illustrated History of North Idaho.

French, Hiram T.
History of Idaho. A Narrative Account of Its Historical Progress, Its People and Its Principal Interests. 3 vols. (1 vol. history, 2 vol. biography) Illustrated.
Chicago and New York, The Lewis Publishing Co., 1914.

A very meager treatment of the Coeur d'Alenes.

Fuller, George W.
History of the Pacific Northwest. Illustrated.
New York, Alfred A. Knopf, 1931.

An excellent and interesting general work, which, of necessity, can give only brief attention to the Coeur d'Alenes.

Fuller, George W.
The Inland Empire of the Pacific Northwest. A History. 4 vols. (3vols. history, 1 vol. biography) Illustrated.
Spokane, Denver, H. G. Linderman, 1928.

Deals largely with Spokane and Washington. Vol. 3, chapter 24 on Mining gives a very meager treatment of the Coeur d'Alenes.

Hawley, James H. (editor)
History of Idaho, the Gem of the Mountains. 4 vols. (1 vol. history, 3 vols. biography) Illustrated.
Chicago, The S. J. Clarke Publishing Co., 1920.

A poor and inadequate history.

Henderson, John M.; Shiach, William S.; Averill, Harry B. (Compilers)
An Illustrated History of North Idaho, Embracing Nez Perces, Idaho, Latah, Kootenai and Shoshone Counties.
Western Historical Publishing Co., 1903.

Part VI, Shoshone County, pages 981–1203. A very complete and valuable work. Contains more documentary material and more authoritative information than any other single work.

Mullan, Captain John, U.S.A.
Report on the Construction of a Military Road from Fort Walla Walla to Fort Benton.
Washington, D. C., Government Printing Office, 1863.

Complete record of early exploration and road building in the Coeur d'Alenes.

Mullan, Captain John, U.S.A.
Miners' and Travelers' Guide to Oregon, Washington, Idaho, Montana and Colorado via the Missouri and Columbia Rivers.
New York, Wm. M. Franklin, 1865.

Good description of primitive conditions in the Coeur d'Alenes.

Onderdonk, James L.
Idaho. Facts and Statistics concerning Its Mining, Farming, Stock-Raising, Lumbering, and Other Resources and Industries.
San Francisco, A. L. Bancroft & Co., 1885.

A very valuable pamphlet (150 pages) written at the time of the Coeur d'Alene gold rush, and therefore containing a great quantity of contemporary material.

Ransome, Frederick Leslie, and Calkins, Frank Cathcart.
The Geology and Ore Deposits of the Coeur d'Alene District Department of the Interior, U.S.G.S., Professional Paper 62.
Washington, D. C., Government Printing Office, 1908.

A very complete study of the geologic structure of the Coeur d'Alenes. Also contains a few historical references and a record of mineral production.

Stoll, William T.
Silver Strike. The True Story of Silver Mining in the Coeur d'Alenes, as told by William T. Stoll to H. W. Whicker.
Boston, Little, Brown and Co., 1932.

An amusing, but recklessly inaccurate volume of reminiscences. Deals chiefly with litigation and labor difficulties on the South Side of the Coeur d'Alenes. Contains almost no documentation and no index.

Umpleby, Joseph B., and Jones, E. L., Jr.
Geology and Ore Deposits of Shoshone County, Idaho. Department of

the Interior, U.S.G.S., Bulletin 732.
Washington, D. C., Government Printing Office, 1923.

A more recent study, similar to, but briefer than the Ransome and Calkins Professional Paper, listed above.

Wardner, James F.
Jim Wardner of Wardner, Idaho, by Himself. Illustrated.
New York, the Anglo-American Publishing Co., 1900.

Interesting personal narrative. Chapters 11, 12, and 18 relate the author's experiences in Murray.

II. NEWSPAPER AND PERIODICAL PUBLICATIONS

Aulbach, Adam
Reminiscences of Prichard Creek, Idaho.
San Francisco, The Mining and Scientific Press, December 20, 1919.

Aulbach, Adam
Coeur d'Alene. Idaho's Famous District Capable of Producing Ten Millions a Year.
Newspaper article dated 1897. Publisher unknown.
(Found in Adam Aulbach's scrap-book)

Aulbach, Adam
The Coeur d'Alene Gold Belt.
Newspaper article dated October 22, 1897. Publisher unknown.
(Adam Aulbach's scrap-book)

Aulbach, Adam
The North Side of the Coeur d'Alenes.
Wallace, Eagles' Souvenir of the Coeur d'Alenes, 1928. Pages 43–45.

Aulbach, Adam
A Snow Bulletin.
Murray, The Coeur d'Alene Sun, January 29, 1913.

Aulbach, Adam
The First Christmas in the Coeur d'Alenes.
Murray, The Coeur d'Alene Sun, 1913.

Aulbach, Adam
History of the Golden Chest, Famous North Side Gold Mine.
Wallace, The Wallace Miner, January 14, 1932.

Aulbach, Adam
High Grading Practices in Days of Gold in Coeur d'Alenes.
Wallace, The Wallace Miner, January 21, 1932.

Aulbach, Adam
High Grading Thrived With Development of Quartz Gold Mines in Coeur d'Alenes.
Wallace, The Wallace Miner, January 28 and February 4, 1932.

Aulbach, Adam
Revival of Placer Mining on North Side.
Wallace, The Wallace Miner, January 21, 1932.

Aulbach, Adam
Murray's First Newspaper Plant Packed Over the Belknap Trail in 1884.
Wallace, The Wallace Miner, February 18, 1932.

Aulbach, Adam
Authentic Story of the Discovery of Gold in the Coeur d'Alenes.
Wallace, The Wallace Miner, February 25, 1932.

Aulbach, Adam
Early Litigation Over Prichard Creek Placer Claims.
Wallace, The Wallace Miner, March 3, 1932.

Aulbach, Adam
Schemes to Work Deep Placers on Prichard Creek and Tributaries.
Wallace, The Wallace Miner, March 10, 1932.

Aulbach, Adam
Rare Picture Specimen of Gold Found in Buckskin Gulch Gives Murray a Thrill.
Wallace, The Wallace Miner, April 28, 1932.

Aulbach, Adam
Veteran Newspaper Man of Murray Tells of the Modern Gold Rush Now on the Historic Prichard Creek.
Wallace, The Wallace Miner, June 23, 1932.

Aulbach, Adam
Placer Mining Activity Along Prichard Creek. Origin and Extent of Rich Ancient Channel Unknown.
Wallace, The Wallace Miner, July 14, 1932.

Culbertson, F. R.
Coeur d'Alene Mining District.
Newspaper article dated 1897. Publisher unknown. (Adam Aulbach's scrap-book).

Rogers, Harry A. (interview)
Discovery of Gold by Prichard in 1881. From County Records.
Wallace, The Wallace Miner, 1925.

INDEX

D

E

F

G

H

I

J

K

L

M

N

S

T

COLOPHON

The Robert Wayne Smith, HISTORY OF PLACER AND QUARTZ GOLD MINING IN THE COEUR D'ALENE DISTRICT, *was printed in the workshop of Glen Adams, which is located in the quiet country village of Fairfield, southern Spokane County and one township removed from the Idaho line. Fairfield is located on State Highway 27 that runs between Opportunity and Tekoa. The paper stock is eighty pound dull coated offset. A limited amount of typesetting was done by Pat Nigh using a Compugraphic Editwriter 7300. Camera/darkroom work was done by Susan Paulson, who also stripped the film and made the printing plates. She used a 20 x 24 inch DS (Japanese) camera. The sheets were printed by Sydney Stevens using a 28 inch Heidelberg press, Model KORS. Folding was by Garry Adams, using a 26 x 40 Baum Dial-O-Matic folding machine. Hardcase binding is by Al and Melissa Chidester, Oakesdale, Washington. Paper binding is by Glen Adams and Garry Adams using a Mark II Sulby adhesive binding machine. This was a fun project. We had no special difficulty with the work. This is the 549th book to come from Ye Galleon Press.*